World War II
Prisoner of War
How I Survived

By Len Kovar

Koho Pono, LLC

World War II Prisoner of War - How I Survived

Published by Koho Pono, LLC
Clackamas, Oregon USA
http://KohoPono.com

Copyright © 2011 Leonard J. Kovar, Carmichael, CA All Rights Reserved

No part of this publication may be reproduced, stored in or introduced into a retrieval system, or transmitted in any form or by any means (electronic, mechanical, photocopying, recording, scanning or otherwise) without the prior written permission of the author or publisher. Requests for permission should be directed to the Permissions Department, Koho Pono, LLC, 15024 SE Pinegrove Loop, Clackamas, Oregon 97015 or online at http://KohoPono.com/permissions.

For general information on our other products, please contact our Customer Service Department within the USA at 503-723-7392 or visit http://KohoPono.com.

First Edition 8september2011

Edited by Jessica Springer

Library of Congress Control Number: 2011933712

ISBN: 978-0-9845424-8-2

Manufactured in the United States of America

"_WWII Prisoner Of War: How I Survived_ tells the story of Len Kovar's experiences as a Prisoner Of War in Germany after his B-24 Heavy Bomber was shot out from under him by enemy fire over Budapest, Hungary during World War II. He tells his compelling story in a sensitive way without dwelling on the horrors of war or the depravity of captivity. It is a book that is hard to put down. Len is an excellent writer!"

- Floyd Mellon, 1st Lieutenant US Army

"I have always had an interest in WWII and have read other war novels. _WWII Prisoner Of War: How I Survived_ is a striking, enthralling, blow-by-blow of a real man's experiences in war. The simplicity of the writing style makes this book an entertaining, easy read - as if it was a grandfather telling his story to friends and family.

Len Kovar is an extremely relatable character - his narration of internal thoughts and struggles demonstrate a battle of fear, courage, worry, and hope that both the young and the old can identify with and understand. This is certainly a book to be shared and enjoyed by people of all ages and experiences.

Len Kovar does an excellent job turning an unfamiliar world into a relatable and identifiable story."

- Kayla Boos, Sophomore, Colorado State University

"To the patriots of America who want to know why we are 'the land of the free' - it is because of the brave like Mr. Kovar. His book, _WWII Prisoner Of War: How I Survived_, not only takes us behind enemy lines as a prisoner of war but reveals to us the war within the war. The interrogations, the intimidation, the endless marching are just the beginning of his story. His trials remind me of Valley Forge type of courage, persistent endurance and will to survive. Mr. Kovar brings us into a world few will ever know much less live to tell about.

These heroes reached deep inside and discovered the secret to self preservation - save each other. Mr. Kovar honestly recounts how he dug deep into his own bank accounts of courage and faith in order to muster a plain, pure will to overcome the imminent life threatening situations.

If you love America, read Mr. Kovar's account of the heroes who were willing to make the ultimate sacrifice in order that we might live free."

- *Michael Nantze, War History Buff*

"Len Kovar writes about being shot down, captured and put into a POW camp during WWII. To escape being captured by the Russians, the Germans forced POW soldiers to march across Germany during the worst winter recorded in the history of Europe. Many died. This is the story of what it takes to survive the equal of the Bataan Death March."

- *Dwight Davis, Military Historian*

"This is an amazing story of survival and perseverance. I thoroughly enjoyed reading this very personal account of one man's journey to return home from World War II after being shot down behind enemy lines. This book reminds me of how much so many people suffer and sacrifice during war. I am so happy I had the opportunity to read this book."

- *Michelle Tambornini*

Dedication

This book is dedicated to my wonderful wife, Lorraine, my faithful companion, who has enriched my life and who has listened to many war stories over the years.

Appreciation also goes to my three children, Tim, Cynthia, and Dave, of whom I am very proud.

Acknowledgements

My warm appreciation goes to Toshia Tolletson whose suggestions and encouragement contributed mightily to the completion of this book.

I also extend my appreciation to Jessica Springer for her insightful editorial help.

I am grateful to my mother, Verna Kovar, who took detailed notes of my story upon my return from Europe.

Thanks also, to the many friends who have encouraged the writing of this book.

- Len Kovar

A Restored B-24 Bomber

Prologue

On August 22, 1944 the 15th Air Force, stationed in Italy, was ordered to conduct an all out bombing mission on the oil facilities and railhead located in Vienna, Austria. Hundreds of aircraft were involved in this mission. We were flying our B-24, named Con Job, in the number four position of the first wave of this assault. This was our eleventh combat mission.

WWII Prisoner of War - How I Survived

It was a beautifully clear day as I sat in Con Job en route toward our mission objective and looked back to see hundreds of bombers flying in orderly formation and stretching back toward the horizon. It was an awesome sight, but I would never see it again.

I was 22 years old, healthy, red-headed and of average height and weight at the start of this mission. Before the war, I had worked for a short time in a brass foundry and then floundered as a freshman in college for a few months. As the war began, I responded to America's call to arms and enlisted in the Army Air Corp. Upon completion of officer training, I became a bombardier/navigator assigned to the 727th Squadron of the 451st Bomb Group. I was now on a major mission against an important target and expected fierce opposition.

Len, February 1944

We were approaching our bomb-drop site at an elevation of 24,700 feet when we were attacked by a squadron of German Folke Wolff 190 fighters. The air space instantly became a writhing, screaming, chaos of aircraft firing twenty millimeter cannon and fifty caliber machine guns. Airplanes exploded into flames. Parachutes trailed out of burning bombers. I saw one parachute burst into flame as it billowed open.

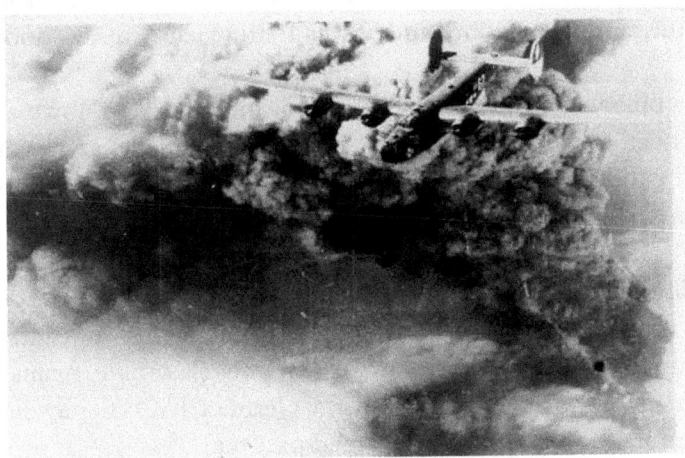

Smoke from Ground Fires and a B-24 Bomber

Con Job jarred to life with the hammering of our fifty caliber machine guns. A German Folke Wolff fighter tore through our formation from our left while another came at us from the front with its cannon firing. Mitch Cohen, our nose gunner, returned fire and I could see our bullets striking the propeller and engine of the incoming German fighter. The fighter exploded and we flew through its smoke, fire and wreckage.

Someone screamed over our intercom, "We're on fire!"

Knowing that the emergency escape hatch for the rest of the crew was through the bomb bays, I hollered back, "Should I open the bomb bay doors?"

"Yes! For God's sake! Open them!" screeched a reply.

I shoved the handle to open the doors and felt the buffeting of wind as they started to grind open. Fire poked its violent fingers around the nose-wheel and reached toward me in my little compartment. I reached up and opened the inner doors of

Cohen's nose-turret gun compartment. He burst out and almost knocked me down. I clipped on my parachute and turned to open the nose wheel doors for our escape. Cohen grabbed for his parachute and watched me intently as I struggled with our doors.

Flames billowed around and grasped for me. I looked down at the earth, nearly 25,000 feet below, and hesitated. I did not want to jump! I looked toward Cohen to seek his advice. His eyes shot forth an urgent command, Go! Jump! Now!

As I looked at the hazy earth far below, an abstract academic question popped into my mind, "I wonder if I'm afraid." Suddenly I felt the violent blow of shrapnel slamming against the heel of my left shoe. The blur of a German FW190 traveling at two hundred miles an hour screamed past me.

Without deciding or thinking I had bailed out of Con Job and pulled my ripcord in one motion. My parachute opened and I found myself amid the massive chaos of a huge dogfight

- Len Kovar

Rick Turnbull found this photo of our B-24 bomber, Con Job, while on our first mission to Ploesti, Romania in 1944. We flew through a heavy flack field and lost one engine.

Chapter 1

It was late morning, about 10 am, on a beautifully clear day. I hung in my parachute at about 20,000 feet and felt the embrace of a profound serenity; a peace that is far beyond the capacity of words to describe. I was being held in the arms of a tranquil silence which has no beginning and no end. It simply is. I gave myself to its infinite embrace.

At about 15,000 feet I awoke from my precious reverie and reluctantly moved my attention to my situation. I had gone from the angry, confused fury of fire-spitting aircraft to the whispering quiet of a vast and tranquil sky. I felt myself to be alone and in the midst of great splendor.

Looking down from my lofty perch, the checkerboard earth was a picture of order, calm and peace. But this was a contradiction to the reality that I faced. This countryside, upon which I would shortly land, was controlled by the enemy: an enemy who would be anxious to capture a terror-flyer like myself who had been bombing their land and devastating their cities. Being apprehended in the midst of a bombing raid was always very dangerous. It often happened that immediate vengeance was delivered on the spot.

I became hyper-alert to my exposed situation. I was alone and nearly a thousand miles behind enemy lines. Even now, the white speck of my parachute and the little black dot that swung below it, were being observed and studied. No doubt many civilians as well as Nazi soldiers were hastening toward my possible landing site. They would be eager to capture and punish such monsters for their awful crimes. This being the case, it was important that I try to guide my parachute to a landing place where I might have the possibility of escape.

The ground below me was made up of orderly rectangular fields. Here and there, tiny rivulets, streams and a few small scattered wooded areas could be seen. It would be safer for me to land in a densely wooded area, if I could manage it, and I eagerly looked for such a place. After a time, I spotted what looked like a small woods that might provide escape possibilities. I decided to try to land there. I jockeyed my shroud lines gingerly and slowly, being careful not to dump the air out of my parachute in the process.

As I descended lower, Lady Luck seemed to take over and I drifted toward a good sized wooded area. This was exactly what I had hoped for. The lower I fell, the faster the woods rose up toward me. The speed of this appeared to increase, until touchdown was only a few hundred feet away. Hastily I prepared for my entry into the trees. I checked the cinch on my helmet, pulled up my gloves, and crossed my legs to offset the possibility of straddling a branch. I waited until my feet were ready to touch the top branches of the trees and then I wrapped my arms over my face and eyes. Cracking branches snapped, popped and whipped around as I crashed through. I felt a hard, sharp pain as my left foot struck the earth. I was whipped up into the air as the tree top holding my parachute jerked me back up.

I bounced up and down several times as the tree adjusted to my weight. Finally, the motion stopped. Looking up, I could see that my parachute was hung up in the top-most branches. Looking down, I saw that my feet dangled about a foot above the grass. I was hanging in mid-air in the midst of a dense wood. Since my feet did not touch the ground, getting out of my parachute harness would be difficult or perhaps impossible. The clasps on my parachute are very tight and could not be unlatched unless it was fully unweighted. This is impossible while hanging in the harness. I twisted and turned and tried to unclasp the parachute but, it quickly became obvious that the clasp could not be released while carrying my weight. Fear welled up within me. I had to get out of the harness soon before enemy captors got to me. But to do this I had to become weightless in order to take pressure off the clips.

I deliberately began to bounce up and down. As I did, the branches bent down and then pulled me back up. Gradually, like a child's backyard swing, the arc began to increase until finally my feet touched the ground at the bottom of each bounce. In that short half-second of being fully-weighted, I attempted to unclip the latches of my parachute. After several fumbling efforts, success came and I dropped in a heap to the ground.

When the tree snapped up to its full upright position, my chute became highly visible in the high branches. My heart, pumping furiously in the tension of the moment, drowned out the quiet of the dense woods. Forcing myself to stand completely still, I listened to the rustling sounds of nature. At first I could only hear normal forest noises; then, from the distance, came the ominous sound of breaking branches. They were perhaps a hundred yards from where my white parachute marked my position. "A German patrol," I thought in panic, "and they are close." They were coming toward me but their progress was slowed by brush and fallen timber. This would inhibit them, but would quickly give way to their determined effort to get to me. I had to leave this place immediately.

It was difficult for me to move rapidly because I was burdened by the high altitude clothing and emergency equipment that I was wearing. This included, among other things, my parachute harness, the May West flotation vest, bulky high altitude sheepskin clothing and boots. Try as I might, it was not possible for me to move rapidly. It was like attempting to swim in molasses. I had to get rid of everything that slowed me down. I tore off my helmet and gloves and threw them to the ground. Hastily, I pulled off my May West, unclasped my parachute harness, and stood still for a moment to listen to the woods.

I heard the distant noises of the crunching and cracking of branches. I couldn't see anyone but I felt sure that people were fast closing in on me. I turned to run in the opposite direction from the sounds, stopped in mid-motion and reversed my direction. I ran toward the sound at an angle of about thirty degrees to the left. After running only a short distance, I saw a large bush with branches that were bent over touching the ground. I dove under the branches, burrowed in as deeply as I could, pulled in my feet behind me and froze.

My heart beat wildly. Each beat seemed to me like the drumming hoofs of a galloping horse. Frantically and with great effort, I tried to be quiet and to breathe slowly. "Relax," I silently screamed to myself in a desperate panic.

My pursuers rapidly entered the area of my parachute. They were fast, efficient and made surprisingly little noise. There were not many people, perhaps three or four, and their whispered conversation was limited. They stayed at the site of my parachute for a brief moment, long enough to decide that they had missed me. Then, they plunged headlong into the woods and, in a few moments, the forest absorbed their blundering noise. I could no longer determine their location but I knew they were out there somewhere doggedly searching for me.

I was close to panic as I tried to lie silently in my hiding place. I was still free from my captors, having done the most illogical thing possible: I had run directly into the enemy and hid at their feet. The searchers would probably not return to my parachute immediately but, they would return. It was clear that my best strategy would be to leave my parachute and this area as quickly as possible. For the moment, however, I was in the best place I could be.

My shielding bush covered me fully but, this was a hot August day and I soon became very warm. My discomfort continued to increase and, in a short time, I became insufferably hot in my heavy sheepskin clothing. I began to sweat profusely. Before long, my discomfort became extreme and I was desperate to get out from under my bush and shed my heavy clothes. That would be impossible until my pursuers were out of the area. For now, I could only do my best to endure.

Mosquitoes discovered my presence: legions and legions, and their reserve armies as well. They attacked me with unrelenting fury. Every patch of exposed flesh was black with them and they were determined to have me for lunch. I was being eaten alive! Never before had I experienced anything close to this. Nevertheless, I needed to be still because any movement could reveal my position. I made an attempt to get rid of them by slowly wiping my face with one hand. As I did this, they would attack the back of that hand. I would scrape them off with my other hand and they would completely cover it. They relentlessly and unceasingly swarmed over my face, head and ears as I sought to fight them off without shaking my protective bush or making noise. The next twenty minutes or so became an experience of intense torture and rank among the longest moments in my life. Gradually the woods became quiet and still. Finally, I came to believe that the area around me was free of people and it was safe to come out.

Exhausted and drenched with sweat, I carefully emerged from my hiding place. I quickly dragged off my heavy sheepskin boots, jacket and trousers, and stripped down to my green flight suit. The relief of being free of its bulk and heat was immense. I spent a brief instant thinking about how much of my clothing I should keep but decided I did not have time for such considerations. I had to get away from this area immediately but, in the months to come, I would regret not having kept several items.

My pursuers had been heading south as they came down the hill in their search for me but, not finding me near my parachute, they continued on. My best escape would be to head north up the hill in the general direction from which they came. They would not expect me to retrace their steps and I would be putting distance between us. That decision made, I stumbled up the hill somewhat to the left of their path.

> *Some 6000 miles away in Minneapolis, Minnesota, while the air battle was going on and I drifted toward the ground, my mother was giving birth to her fourth and last child, my sister Susan. The birth process was normal and not unduly rigorous. The child was healthy and strong. In accord with hospital procedures, the baby was taken to the nursery to be bathed and cared for while my mother was moved to the recovery room, a quiet room with subdued lighting. She was resting comfortably and beginning to come out of the effects of the twilight sleep medications used during delivery. Suddenly, she was startled as she saw me standing at the foot of her bed. She saw me as though I was physically in the room with her. She described me as wearing a green jumpsuit, I was sweaty, my hair was rumpled, and I looked worried and afraid. Her vision lasted for several minutes and then gradually faded away.*
>
> *She knew that, even though she saw me as physically real, I was not actually in the room with her. The fact that she saw me clearly and life-like, troubled her. She wondered why this vision should come to her since both she and the baby were healthy and well. What she had seen however, was so unusual and graphic that she felt the need to share it with someone. She called in a nurse and told her the details of the vision. The nurse listened sympathetically, made a comforting remark, and promptly dismissed the whole event. My mother, however, continued to be bewildered and upset. She wondered why she would see me worried and afraid.*

At that moment in time, a continent away, I was dressed in the long green flight clothes that she described. I was sweaty, my hair was rumpled, and I was worried and afraid. I was on the ground well behind enemy lines, being hunted in a heavy forest, by those who wished me anything but well.

As I proceeded, I noticed that the nature of the woods began to change abruptly. The dense woods became clear of brush and forest debris and became something akin to our American parks:

clean and manicured. The trees were plentiful but few of them had branches below head level.[1] The further I went into these woods, the more cultivated the woods became and the less cover I had. I was becoming increasingly concerned as the danger of being seen grew. It was urgent that I find a safe, out-of-sight, hiding place soon. Time was running out. Even now search parties could be forming to comb the area.

I stopped to assess my options and to catch my breath. Although there were trees in all directions, the terrain became increasingly open as it led to a hill which crested several hundred yards ahead. From the top of the hill I would be able see in all directions and, from there, I could better decide what to do next. Of course I could also be easily seen from that exposed point but, that was a risk I would have to take.

Walking carefully, I made my way to the top of the rise where I had an excellent view in all directions. The woods appeared to be quiet and serene; nevertheless, I needed a safe place in which to hide before some sort of patrol began its search. As I looked around, I noticed a depression in the grass on which I stood where an animal had bedded down. I decided that this spot on the top of a hill might provide a good hiding place. First, this was so unlikely a hiding place that no one would expect it to be used by a fugitive. It was much too visible from all directions. Furthermore, it would probably be free from traffic because people do not generally climb a hill if there is an easier way around.

---

[1] The reason for this, I learned later, is that the Hungarians had learned long ago to conserve wood and timber. They avoided cutting down trees for firewood whenever possible. Instead they would pick up the fallen wood and when possible cut only the lower branches. This would allow a forest to produce firewood for many years. The result was something akin to a heavily wooded, but carefully maintained park.

I lay down on my back in the matted area to check it out. In doing so, I discovered that the foot long grass in the area extended above me by several inches and hid me quite well. I reasoned that if I stayed motionless, so that the grass did not move, I would not been seen. This illogical spot was as good a place as I might find. With that, I stretched out in the matted area and tried to disappear into the stillness.

Slowly the sun rose to its zenith. Once again I became very warm and began to sweat profusely. In a short time my discomfort graduated to a minor misery and, as hoards of mosquitoes arrived, my misery rose toward anguish. The mosquitoes attacked me with a vengeance. I fought them as best I could.

In the distance, I thought I could hear people signaling to one another and the snap of breaking twigs. These sounds seemed to be gradually surrounding me. The manicured woods were being carefully searched and I was the quarry they sought. My only choice was to lie absolutely still and make sure that I did not reveal my position, even as the mosquitoes continued to swarm over me. Finally, after what seemed an eternity, the sounds abated. The woods took on the quiet of a forest at peace.

I stayed hidden until late in the afternoon before I decided to risk sitting up to look around. Slowly and carefully I rose up on my elbows and scanned the woods. Everything was quiet. Only the buzzing of insects could be heard. I was exceedingly thirsty and also miserable from mosquito bites but, I was still undiscovered. I knew that the search for me would continue but I had made it past the first big hurdle. Relief surged through me along with a bit of euphoria.

I interpreted my having evaded capture as a sign that I would somehow continue to do so. I would find my way back across several countries and time zones and return to Allied territory. I became cocky, elated, and even manic. Of course I needed a plan but, I knew I would find my way back. The luck of the gods was with me.

My first act would be to wait where I was until dark, find water, get my bearings, and set out toward northern Italy. I would be careful and smart, and the seas would part before me as I came to them. While I waited for the lengthening shadows of dusk to appear, I carefully inventoried my equipment. I did not have a jacket but it was August. I had a good hunting knife which had been carefully crafted for me by my father who was a skilled tool and die maker. He had proudly declared that it had been properly tempered and was as fine a knife as could possibly be made. I had a small, basic escape kit in one of my leg pockets which contained matches, a compass, a rubber water bag with a draw-string at the top, a colored map of the Baltic countries printed on a piece of silk cloth, water purification tablets, fishing line and hooks. I carried a small New Testament in my shirt pocket. In my pants pocket I carried a pocketknife, my wallet containing $107.^{00}, several pictures of family members, a driver's license and several other forms of identification. I went through all of my things, then dug a small hole and buried the few papers that might be of value to the enemy should I be captured.

I took out my New Testament, opened it at random, put my finger on a verse, and read it carefully. I don't recall which verse I picked out, but I do recall working at it until it suggested what I wanted it to say: namely, that I would escape capture and find a way to return to Allied lines.

The shadows of imminent darkness were beginning to lengthen. It was time to move on. I studied my silk map and reviewed my plan: I would find water, then find a road and begin making my way toward Yugoslavia several hundred miles to the Southwest. Once there, I would try to find the air field where escaping airmen might be picked up and returned to Allied control. Meanwhile, I would avoid people, travel at night and live off the land.

My secondary plan, if I could not find help when I got to Yugoslavia, would be to make my way across Austria, the northern part of Italy. I would make my way down the East side of German occupied France and across the Pyrenees Mountains and into Spain. Once in Spain, where there was a friendly government, I would turn myself into the authorities. Looking back, neither of these plans was remotely possible, especially in wartime when everyone and everything was under scrutiny. But one has to start somewhere and youth does not always know what cannot be done. Moreover, this was a genuine all out war and being captured by the enemy carried with it the likelihood of being beaten and tortured, the certainty of imprisonment, and the very real possibility of death. It was a time for impossible plans.

I was sitting quietly, studying my silk map when I heard the snap of a twig. Instinctively I froze in position. Without moving my head, I lifted my eyes and saw a large, muscular, middle-aged man laboring his way up the hill directly toward my hiding place. He was close, only thirty or forty yards away. He was dressed in worn work clothes and was obviously tired. He walked absent-mindedly along the faint hint of a trail that passed within one yard of me. I remained motionless as the man walked toward me. But the man's mind was not attentive to his surroundings. He was in his own inner world and he looked at the outer world with unseeing eyes. I looked directly at him and, for a split second, our eyes met but, he neither flinched nor broke his pace. He never saw me although he passed by within three feet of where I sat. He continued to labor his way past me and walked on down the hill. When the crunch of his footsteps became distant, I carefully turned my head and watched him disappear into the trees. I began to breathe again.

Chapter 2

The evening shadows lengthened, as did my sense of isolation and vulnerability. The reality of my circumstances was beginning to register. I was truly alone, hundreds of miles behind enemy lines, without equipment, supplies or friends. I had avoided capture thus far but, clearly it was important that I move out of this area and find a safer hiding place and water.

It is said that one can live with hunger for a long time and still be rational and maintain one's dignity, but thirst is a completely different matter. This ancient observation was becoming a personal fact for me. At this point I had been without water for about sixteen hours. During that period I had been in an open aircraft, at high altitude, for about four hours which is dehydrating. Another ten or twelve hours had been spent sweating in sheepskin clothing under a bush or in the sun. My need for water was now becoming an acute problem. My immediate needs were simple and clear: I must find water, a safe hiding place and, if possible, some food. I would then figure out a way to get back home.

I left my exposed little knoll and started to walk in the general direction from which the large man had come. I tried to avoid the vague path that the man had used because it probably led to a place where there were people, and people were the one thing I had to avoid. I walked cautiously and steadily into the woods hoping to find a road or trail, or better still, a little creek or pool of water.

The evening shadows continued to deepen and become ever more deceptive and impenetrable. Before long, a heavy darkness blanketed the forest. All shapes began to meld together into a general confusion. I began to find it difficult to distinguish between trees, brush, earth and sky. Shock was beginning to set in and my body was functioning in strange ways. My eyes were not working properly, my left foot ached, my perceptions were unclear, I was profoundly tired and I was in desperate need of water. It was important that I get away from this dangerous area and find a safe place to rest and recuperate. Moving along with as much caution as I could, I noticed that the forest was again becoming tangled with brush and forest debris. Progress was becoming more difficult.

I blundered along through the heavy brush and dense woods. Suddenly, and out of nowhere, a small cottage appeared a few yards ahead of me. Startled that I had not seen it sooner, I stopped in my tracks and stood still to get my bearings. It was very quiet. No noises or lights disrupted the darkness. After studying the house for a moment, I decided that I was standing at the back corner of the house. I thought I could see rain gutters coming from the roof and windows on the side. I was startled: I had almost stumbled into a home in the middle of this heavy forest. There were no signs of activity but it was obvious that I had to get away from this area immediately.

Keeping the house on my left, I carefully and quietly walked next to it when I stumbled and fell into a pile of brush. As I picked myself up, I realized that I had been seeing things that were not there. No house existed next to me, only brush and shadows. I was hallucinating: my senses were not dependable. I was faltering along only half functional. After climbing up a sharp rise, I finally found myself standing on a well-kept gravel road. A sense of relief flooded over me. I was somewhere.

This small back country road was perfect since it would now make travel easier for me. Moreover, a gravel road, in the middle of nowhere like this, would have very little traffic. It would therefore, be easier for me to avoid people. I was in luck. I could now begin my hike toward Allied lines and, along the way, I would find the water and shelter that I needed. The road ran north and south. I decided to walk toward the North. The moon came out and I could see the dark silhouette of tree-tops on each side of the road. My vision gradually improved and I began to trust my senses a little more. I became aware of the delicious, refreshing and wonderful smell of fresh water. It was ahead of me a hundred yards or so. I knew that a horse can smell water from a long distance and can sense about how far that water is. I now experienced that for myself and it was very real. I quickened my pace and eagerly walked toward this refreshing aroma which grew stronger as I progressed toward it.

Following my nose down into the ditch, I found what I hoped would be a good sized puddle of water. I eagerly bent over this delectable pool of life to find that it was now only an area of mud covered with a thin film of water on the top. I dug a hole into the mud expecting to gather a pool of water from which to drink, but, my labors only succeeded in creating more mud. After several frustrating tries, I realized that the water I craved would not be found here. I would have to move on.

As I hiked down the road, I seemed to recover somewhat from my fatigue and began to get my second wind. With my eyesight improved, I was beginning to feel stronger and more alert. I continued down the road for another mile or so and again I smelled water. The odor was strong, delicious, clear, and close by. A short distance later, I came to an open driveway that led to a large two-story farmhouse located about fifty yards in from the road. This farm was obviously occupied. Various pieces of equipment were strewn about and in the front of the house was a well. I would have to be speedy about what I did. I decided that my best strategy would be to enter the yard, quickly, draw up a bucket of water, take a good drink, fill my little rubber water container and be gone.

Cautiously I crept into the yard and made my way toward the well. About half-way to the well, a dog began to bark from within the house. Damn it! A dog! Maybe he will quit barking, I thought, although I knew better. I continued to move forward toward the water when a second story window banged open. A large man stuck his head out of the window and looked around. The moonlight was bright and I could see that he was wearing a nightcap as well as a nightshirt. I froze in my tracks. The man surveyed his yard for a moment. The dog barked again and the man abruptly tucked his head back inside. The man must have suspected that someone was in his yard and he was coming out to investigate. It was time to leave and quickly. I hastily retreated into darkness and back to the safety of the road. I would have to find water somewhere else.

As I made my way up the road, a thin cloud cover started to form and the moonlight gradually became diffused and dim. It became difficult to see the road or even the silhouette of the trees. The night had become still and quiet. I could hear few sounds except for the occasional wail of a loon and the scrape of my shoes on the gravel. I felt alone and uncertain.

I hiked on for quite awhile and then became aware of the sound of footsteps coming toward me from the road ahead. The footsteps were heavy and steady. It was a man and he was alone. Anxiety along with indecision welled up within me. How should I handle our meeting? Neighboring people meeting in the middle of the night would certainly stop and talk to one another. There was no way that I could do that without giving myself away as an American who did not belong there. If I was discovered as an enemy soldier what should I do? Try to fight him and get away? If I fought him, should I try to kill him? If I did kill him, where would I go and what would I do next? My inner conflict continued and became more confusing as we drew closer to each other. What to do?

If I heard the man coming, then he likely heard me. The local people would certainly be aware that downed American airmen might be in the area. If I were to leave the road now and try to hide, it would be obvious that someone was trying to avoid recognition. If I said nothing and continued to walk toward him, we would pass within an arm's reach of one another. If he spoke to me, I could not respond because that would reveal that I did not belong here. That could be disastrous. My inner debate continued as we marched steadily toward each other on a dark, lonely, eerie night.

He was close now, close enough to reach out and touch. Without speaking we crunched past each other in silence. I headed north on my right side of the road and he marched south on his right side of the road. It occurred to me much later, that in Europe people normally travel on the left side of the road. In America we normally travel on the right. On this occasion, we were both walking on our right. Neither of us spoke on that dark night. Was it because we were both Americans and afraid of being recognized? I wonder how the war might have turned out for me if we had recognized one another and joined forces.

The silhouette of trees on each side of the road had given way to an occasional open pasture or cornfield. The landscape had changed and again I could smell water. The road turned at a dogleg and before me was a small village no more than four or five blocks long. Houses lined each side of this main road as it passed through the town and other dwellings branched off on both sides. I entered the village carefully. It probably housed working farm people who were now asleep. I was in luck.

The smell of fresh water was strong and I moved toward it quickly. I assumed that the scent was coming from a well located at the center of the town. As I proceeded, I saw the carefully crafted stone wall that surrounded the actual hole. Next to the stone wall was a long pole with a heavy weight attached at one end. The other end had a bucket attached to it with a rope. To draw water one had only to lower the bucket into the well, let it fill with water, then lift it out. The precious elixir of life was close at hand.

I was less than a block away from the well when I heard a dog bark. A moment later, a dog trotted warily out from a nearby house. He looked uneasy and menacing as he edged his way toward me, the hair on his back raised. Fear rose within me. This was a nervous dog who knew that I did not belong and he was ready to defend his territory. Another dog emerged from shadows as I continued on. I became even more fearful when two or three more dogs gathered and circled around me. They were tense and uneasy: a stranger with a different smell was in their midst. Their natural protective instincts rose up. They surrounded me with bristling hair, curled lips, and uneasy prancing. My throat tensed, my legs stiffened and it became difficult to move my legs. Barely able to walk, I tried to control my fear, but in spite of my strongest effort, I was nearly overpowered. Terror was taking charge: a primitive instinct within me had remembered that there is no killing machine in the natural world as efficient as a pack of dogs. But I absolutely had to keep on walking: I sensed that if I stopped they would attack.

Nearly paralyzed, I managed to force myself forward. More dogs joined the pack until there were about eight or ten nervously milling about with uncertainty. This harrowing procession continued until I reached the end of the village. Then, one by one, they began to disappear into the night. Finally I was alone. I continued to push myself out of town. I had just lived through a level of fear that was extreme. Now limp, exhausted, and barely able to stand, I sat down and tried to recover my courage, my strength and my wits.

I never before realized that there are many levels of fear. Flying into the oily splotches of exploding anti-aircraft shells evokes one kind of fear. Bailing out of a burning bomber at 24,000 feet in the midst of a huge air battle raised that fear a notch or two. These fears are real and deep, but they are recent, modern day fears and they only go back a few generations. The instinctive fear of being torn to shreds by the gleaming teeth of a dog pack is different; it is a primal fear and dates back to antiquity.

The night was well along and I had put space between myself and my parachute, but I still needed water, food, rest, and a safe hiding place. Dawn would soon banish the protective cover of night. Again on the move, I approached another small town that was a bit larger than the previous one. Once again my nose drew me toward the town well. I moved purposefully, anxiously hoping that I could successfully draw water, drink my fill, and get out of town without being noticed. This would be my last chance before daylight.

The streets were empty as I entered the village. Down the road I could see the town center and community well. Arriving at the well, I quickly started the process of drawing water. As I moved the pole into its proper position, I felt the pressure of eyes examining me from behind. I slowly turned toward the presence that studied me. There, at the far end of the town, I could see a man standing and studying me in the half-light of a struggling early dawn. He looked suspicious. I had to leave now. I stood up and stretched in a leisurely manner, pretending that I belonged to this place. Then I headed on toward the end of town.

The road continued for a couple of blocks then turned left at a dogleg. There, two large wrought-iron doors stood open and led into the yard of a home. I again smelled water coming from a well inside the gate. I quietly entered and approached the well. By now I had become familiar with how this kind of well worked. The bucket was handy and I lowered it into the water. I quickly filled the bucket and started to draw it up but the process was noisy. The bucket leaked and the sound of water falling back into the well amplified as it reverberated and bounced about. I needed silence.

Once again I could feel the presence of someone coming toward me. A certainty rose within me and I knew that I must leave immediately. I hastily placed the half-full water bucket on the grass where it would not drip. I quickly filled my little rubber water sack, retraced my steps through the gates and turned left to get out of town. I walked about a block to the end of town and then continued another half mile. There, I came upon heavy woods that bordered a large cornfield. The six foot high corn was healthy, tall and ripe for the harvest. Between the woods and the field was an old hedge. It was thick and wide and its branches hung over and touched the ground. I walked along the hedge for a short distance and discovered a large and neglected bush which looked to me like the protective cover I needed. I burrowed under the lower branches and found the ground to be fairly flat and clear of debris. Further, it was a large enough space in which to lie down and it was completely hidden from the outside. This, I decided, was about as ideal a hiding place as I could hope to find.

I crawled into this little safe haven and started to prepare it as a place to sleep. Reaching up the side of the trunk, I felt the stump of a branch that protruded out about an inch. This branch could hold my rubber bag with its precious water. I carefully placed the strings of my water bag over the stump to hold it while I sat down and prepared the space so I could rest and drink the water. I began to brush away the stones and twigs when I suddenly found my hand wet with mud. Where did the mud come from? I reached down again and found my water bag – empty. The small branch on which I had placed it had given way. I was stunned and bewildered. My thirst was desperate but my water was gone. I had no choice now but to accept this fact, stay where I was and wait for nightfall to try again. Barely able to function, I stretched out in my little haven and collapsed into a deep sleep.

Chapter 3

It was mid-afternoon on a lovely day when I awoke with a start. I was fully alert. There were people close by. I could feel it and sense it. I sat up, hardly breathing, and listened. I heard vague and unfamiliar noises that were fairly close. I sat still and attentive. Within a moment I located the sounds: they were coming from the cornfield on the other side of the hedge adjacent to my hiding place.

After listening for a moment I determined that I was hearing the voices and activities of women. They sounded as if they were in a friendly and cooperative mood. I concluded that they were picking corn together. I had nothing to fear from them as long as I was quiet and stayed out of sight.

I was in pretty good shape. I had slept nearly eight hours and recovered a good deal of my strength. Despite a dry mouth with a powerful thirst, a nagging hunger and a sore foot, I was in good shape. My next move was clear, I would stay hidden until after dark. Then, when all was quiet, I would return to the well at the edge of town. There I would fill up on water and then strike out for Yugoslavia, hundreds of miles to the Southwest.

The work party departed the cornfield in late afternoon. After their voices faded away in the distance, I cautiously crawled out from under my hiding place and explored the secluded area. I found a patch of blackberries but, since it was past the season for fruit, I found very few. I pulled off an ear of corn and chewed it in an effort to suck out moisture, but, raw corn provides very little liquid. Instead of helping to reduce my thirst, it puckered my mouth and increased my need. It became clear that I would have to put up with my dry mouth and demanding thirst. There would be no relief until I could return to the well in town under the cover of darkness.

The sun continued its journey across the sky and as it dipped into the horizon, my need for water continued to grow. Impelled by a dry mouth and cracked lips, I began to consider approaching the well, not at midnight, but earlier in the evening. The thirstier I became the better that idea seemed to become. By the time the shadows of dusk appeared, I had convinced myself that going into town before midnight, was a good idea. After all, I reasoned, most of these hardworking farm people would be going to bed shortly after dark. Few people would be around at that time and I would be able to avoid them.

Dusk and its deepening shadows finally began to appear. The last strands of light reluctantly gave way to the insistent approach of dark. It was nearly nine at night when I left my hiding place and secreted my way toward the well at the edge of town. The village appeared to be quiet as I arrived at the boardwalk that ran beside the concrete wall. I knew there was a large iron gate at the other end of the wall and, just inside the gate, would be clear, fresh, delicious water.

I cautiously edged my way toward the gate when suddenly I heard footsteps coming from the village center. I sensed that what I heard came from a young man and woman who were taking a nightly stroll together. My intuition heightened, I sensed that they were going to turn at the dogleg. They would then walk directly into me. I knew I had to get off of the boardwalk and hide immediately.

I anxiously looked for an escape and then noticed an old wooden doorway located in the wall next to me. This could be my escape. I turned the latch on the door and pushed, and the door opened. Quickly and silently I let myself inside and closed the door. The man and woman were involved with one another and oblivious to my presence. I held my breath and stood behind the door as they passed by me only inches away.

I waited as their footsteps faded into the distance. Then, looking around, I discovered that I was standing next to a small building. It reminded me of an old-time icehouse or tool shed. I could see that I was standing at the edge of a large, well-kept yard. An old church stood at the end of the yard. On the other end of the yard, near the gate, was the well with its precious water. Set back a short ways was a large old house, probably the church manse.

I quickly made my way to the well, lifted the long pole with the bucket attached and gently lowered it. The bucket touched the water and began to fill. When the bucket was about half full, I started to raise it up toward the surface. But, the bucket leaked and the sound of leaking water falling into the well splashed up noisily. The sound reverberated against the walls of the well like it was a sound chamber, amplifying the noise as it rose. It became a clanging bell announcing my presence. Just then, the iron gate at the entrance squeaked. I looked up, startled. Someone was entering the yard. Both the noisy bucket and I needed to be silent and out of sight immediately. The bucket was too near the surface to lower it to the bottom quickly, so I jerked it up out of the well. I placed the bucket on the ground so that the dripping could not be heard. Then I ducked out of sight behind the well.

The small man coming through the gate was dressed in civilian clothes. He briskly walked two or three steps into the yard, then, paused and looked around for a moment and listened suspiciously. Then, deliberately closing the gate, he turned and walked purposefully to the front door of his house and quickly entered. I thought he had sensed that a stranger was hiding in his yard. Moreover, he had a very good idea of where that stranger was located.

It was clear to me that the man would be coming out of the house soon and he would be alert and on guard. A debate clashed within me. Should I run up to his front door and attack him as he came out of his house? But to attack and perhaps kill a man - in his hometown - while still hundreds of miles inside enemy territory? The idea was not workable. Moreover, I did not have the stomach for it.

I turned to leave immediately. The iron gate squeaked as I squeezed through. The young man and his girl-friend were on the street behind me so I had no choice but to turn right and head toward the center of town. I crossed the street and walked on the other side of the road for a short block and then turned a corner and entered a side street. I glanced back toward the gate and saw two men turn the corner behind me. They were following me. I increased my pace and, a short distance further, crossed the street again. Glancing back I saw that the two men also crossed the street behind me. They were following me and gradually catching up. I crossed the street again and they did the same, but, now they were considerably closer. Another block farther and they came abreast, with one on each side of me.

The three of us walked together side by side for a few moments. No one spoke. I smiled at one man who smiled back. I smiled at the other, who also smiled back. I debated with myself about what to do. Should I attack both of these men and hope to overpower them? They were both small men and I was in good shape, maybe I could take on both of them - and win? Possibly not. Most likely they were armed. Moreover, even if I did defeat them both, what then? I was still deep in enemy territory with no equipment or friends. We walked silently for a minute or two as I continued my inner debate. They smiled at me and I smiled at them. Finally, I blurted out: "O.K. Jack, I'm an American."

Immediately everything changed. Acting firmly and deliberately, as though they knew exactly what to do, each man grabbed one of my arms and briskly steered me off the boardwalk. We quickly traversed an empty yard, an alley, and across a couple of unkempt lots. Within a few moments we approached a large, old, sprawling, single-story frame building. The building looked like an out-of-use dance hall or community center that was now decrepit and forlorn. Hustling me up the three step entrance, they opened the door and pushed me in.

Chapter 4

The room was dark and had the hollow ring of neglect and infrequent use. Black draperies covered the windows in keeping with war-time black-out restrictions. The hollow sounding room was vacant except for one long table and a few chairs placed along one side. At the front of the room was a low stage that was separated from the larger public area by a low railing. A small gate at the center allowed access to the stage.

I felt apprehensive as they walked me through the gate and onto the stage, closing the gate behind me. One of the men held me firmly while the other secured a straight back chair and indicated that I should sit down, which I did. They then stepped back and stood guard.

My situation was clear: I was being held by civilians who were businesslike and efficient. They had some sort of organizational procedure and it appeared to be unfolding properly. They did not plan to harm me as long as I did not resist or cause trouble. They had a job to do and they knew how to do it. The question that troubled me was: who were they? Were these people Partisans who might befriend me, or, were they Nazi supporters who might turn me over to the Gestapo?

I sat in my chair quietly for a time until the front door opened and a very large, middle-aged man entered the room. This man exuded the confidence of authority. He was accompanied by three other men. Hardly bothering to glance at me, the men walked over to my right where the big man sat down at the long table. He gave an order to one of his associates, who immediately left the room. A few moments later, three or four more men entered the room and gathered at the table to discuss my fate. Two of the men were

WWII Prisoner of War - How I Survived

ordered to search me. They relieved me of my hunting knife, escape kit, wallet, and New Testament, but, they missed the little pen knife in my pocket. After gathering my belongings, the men brought them to the table where the big man examined the items. He carefully inspected my identification papers and ordered one of his men to check my dog tags.

Gesturing to me, the big man made it known that he wanted to know my name, rank and serial number. I responded for all to hear, "Leonard J. Kovar, 2nd Lieutenant, 0706317." He grunted his acceptance of this report and finished his examination of my meager supply. He ordered one of his assistants to wrap my things in a package. Meanwhile, several more men came into the room to take part in the proceedings, until there were about a dozen men present.

The group of men entered into an intense and somewhat heated conversation. I had no idea of what they were talking about, but, I assumed it had to do with what they should do with me. As the men debated, a large, buxom, middle-aged woman entered from behind the stage carrying a pitcher of water and a plate of food. One of my captors took the tray from her and set it down near me and indicated that I eat and drink. Eagerly, I poured water into the glass and gulped it down. It was clear, clean, delicious, refreshing and wonderful. I poured still another glass of water and gulped it down. As I poured another, one of my captors stepped up, held my arm and pointed to the food. I wanted the water but he pointed to the food. He knew that I was dehydrated and should not drink too much too soon. He took the glass from me and held it so that he could dole out water to me at his discretion.

The food on the plate consisted of a piece of green pepper and a good-sized piece of ham with a rim of fat about an inch thick. I ate the green pepper and the ham leaving only the fat. I still had American eating habits and did not have the palate for straight fat. In the months to come I would have given a great deal for that rich and delicious piece of dripping pork fat. I wanted more water and, during the next hour, they doled it out to me gradually.

The intense conversation between the big man and his aides continued. An old man, about seventy or so, bundled in warm clothing, was respectfully escorted into the room. He was gently assisted to a chair that had been placed on the other side of the rail fence from me. Apparently he had been brought in as an interpreter. I came to understand that he had visited Boston many years before and spoke some English. As it turned out, his English was extremely limited and my Hungarian was zero. We tried to converse but it was difficult; however, I did finally grasp what I thought was his question, "did I want to go to Turkey or to Norway"?

I was elated, almost euphoric, at my good luck. I was with Partisans, who would get me back to Allied lines. As the old man and I attempted to communicate, the heated discussion continued among the men around the table. I assumed that they were discussing the various options and details involved in possibly returning me to my home base. Let them figure it out, I thought. When they say 'jump', I will only ask, 'how high'? I would be totally cooperative.

I eagerly tried to converse with the old man. We puzzled with one another for a time, until finally, and, with considerable effort, I understood him to warn me that I must follow directions exactly or I would be shot. I immediately agreed. I knew that total cooperation was the absolute, unbending rule of the underground. I tried to convey to him that they would have my complete and unquestioning cooperation.

I felt good about how things were going. I had been fed and my thirst had been abated. I sat back and reached into my breast pocket for a cigarette. Suddenly, the room became electric. The conversation stopped. The big man dropped his hand into the drawer in the table. Others shifted position. All eyes focused on me. Silence gripped the air. I froze in place with my hand in mid-air. Something was wrong. Time stood still.

I realized that several guns were pointed at me. They must have thought I was reaching for a gun and were ready to shoot me. Very slowly and with great care, I drew my hand out of my flight suit and lifted my open hands over my head to show that they were empty. I anxiously smiled at all in the room to show that I was friendly and cooperative. The room began to relax and the conversation resumed.

The old man motioned for me to get closer to him. Very earnestly he tried to communicate with me. Finally, and with effort, I understood him to say something like, "they thought you were going after a gun. You must be very careful and not do things like that. Your name is Kovar. This is a Hungarian name; therefore, you are a traitor to our country and you should be shot right now. Others think that you should be turned over to the German Gestapo. They are now trying to decide what to do with you. It is very possible that you will be shot. I advise you not to provoke anyone." He drew his hand across his beltline indicating that half of me was good, Hungarian, and the other half was bad, a traitor to the Hungarian cause.

I was shocked. I went from euphoria to sheer terror in an instant and it showed in my face and behavior. They were discussing my fate. Their heated discussion did not have to do with how to help me escape and return to my home base but, rather, should they execute me as a traitor or turn me over to the dreaded German Gestapo?

I tried to control the surge of panic that rose within me. I turned toward the old man intending to beg for my life. I started to kneel before him and, as I did, a dark scowl of contempt clouded his face. His expression was eloquent and clear: Such cowardice, a soldier begging for his life and a Hungarian at that; disgusting.

I was halfway down on one knee when I suddenly realized that such fawning and cowardly activity worked to my disadvantage. I caught myself before my knee hit the floor and reversed my direction. I stood up straight and tall and I looked at the old man and the entire room with the cold eyes of resolve. I tried to convey the thought: To hell with all of you; do what you will, you bastards!

I paced up and down in my confined area for a moment. Then I sat down, crossed my arms and looked across the room with an air of firm resolve. This may sound like bravery on my part but it was not. It was an involuntary act. It was as though some hidden part of my unconscious took over and lived me for that moment and produced what was needed at the time. In a very real sense, the objective person known as Len Kovar was a by-stander, an observer of this event. I did not do the reacting so much as something deep within reacted me. In retrospect, I must say that my response was exactly the right thing to do at that moment to salvage the situation. This event happened in spite of me and not through any act of bravado, courage or cleverness on my part.

The conversation among the men continued for quite some time. All the while, I sat stoic, impassive and controlled. Finally, they came to a decision. The big man issued an order. With that, one man rose and left the room while the other men began to relax and mill around. Shortly, the man with the orders reentered the room followed by two young boys about sixteen years old. Both of these boys were wearing homemade military hats with homemade insignias attached and each carried a rifle. It appeared to me that their rifles were single shot 22's and reminded me of the kind of rifle that my buddies and I had used to shoot gophers in the Minnesota woods just a few years earlier. These boys looked dangerous to me, as though they would be proud and eager to do something heroic. I felt wary of them yet I was relieved. Something was being prepared for me and it would not be execution. An execution would not be assigned to a couple of young boys. Men came forward from the group to escort me outside and with that, the meeting broke up.

The moon was partially hidden and it was quite dark on the dirt road at the back of the building. A dozen or so sullen and suspicious townspeople had gathered to observe. They were standing near a horse drawn buggy which, in an earlier generation, had been an expensive carriage. Seated in the front seat of the carriage was an old man, who appeared to be about eighty years old, wearing a warm cap and a heavy coat. He was holding the reins to a single horse and he appeared to be about half-asleep and unconcerned about what was going on around him.

I was directed to get into the front seat of the buggy next to the old man. The two boys swaggered into the back seat to sit behind me. They were very proud of their responsibility to guard this enemy in their midst and this evoked an uneasy feeling within me. The big man in charge approached me and earnestly tried to make me understand that I should not attempt to escape. He made it clear that the boys were ordered to shoot me if I tried.

The old man shook the reins and the horse began to lumber forward into the night. The buggy turned onto a little side road that led out of town. The old man appeared to have nodded off to sleep and merely held the reins as the old horse plodded on. The boys in the seat immediately behind me were enjoying their authority and power. One of them stuck his rifle very close to my side and fired at a tree on the side of the road. While he reloaded, the other boy picked out a target and, aiming so that he would just barely miss me, fired his weapon. It was great sport and the boys enjoyed it immensely. The old man dozed through it all. I aged considerably.

This entertainment continued until the boy behind me discovered a new dimension to this sport. He would accidentally jockey his gun until the barrel pressed firmly into my back. There it would bump and bounce against my back as the horse lumbered along. I knew that this was only to frighten me. They did not really intend to shoot me but it was unnerving nevertheless. Whenever one of them pushed his gun into the center of my back, I would turn around slowly, very slowly, so as not to alarm him, and gently push the gun away. The boys continued this activity for quite awhile.

We passed by a narrow stream next to a picturesque little church. I wanted to divert the boys' attention so I pointed to it and said, "church." One of the boys responded by telling me the Hungarian name which sounded something like Kirk. I tried to get them to tell me the names of things, anything to distract them from their sport of shoving a rifle into my back or shooting just past my side. The old man seemed oblivious to all of this as he dozed in his seat and the horse dutifully plodded along the dark, but familiar, road.

We finally arrived at the edge of a small village where the old man brought the horse to a stop adjacent to a large, old, stone building. While only a few hours had passed, it felt more like centuries to me. The two boys, now acting very businesslike and efficient, ordered me out of the buggy and directed me to a door in the side of the building. It appeared to be an old military barracks. We entered a small, cluttered, dimly lit office room. A radio was softly playing. The room was dominated by a desk that was littered high with miscellaneous items including: American money, identification cards and personal items that appeared to have been taken from prisoners before me.

Seated behind the desk was the officer on duty. He appeared to be a strong, competent Hungarian soldier about thirty-five years old. He looked both bored and tired. Behind him another man, a bit older, was sleeping in a straight-back chair. The young boys pushed me into the room. Feeling very important, they clicked their heels and saluted the officer on duty in a smart military fashion. They delivered the wrapped package that contained my possessions. They spoke to him briefly and then, saluting smartly, they left the room.

The soldier on duty looked at me without emotion. He was tired and was simply doing his job. Whatever was about to occur, he had done before. The noise of my arrival awakened the older man who had been sleeping in the chair. He joined in the process of looking through the package containing my things and checking to see if I was armed. My possessions were tossed onto the desk along with the mound of previously collected items. The older man returned to his chair in an attempt to go back to sleep. The soldier on duty pointed to the floor, put his hands together next to his cheek, indicating that I should lie down and go to sleep.

I looked at the floor and noticed a large object in the shadows under his desk. The object looked strange and I could not quite make out what it was. Suddenly, I realized that it must be a man lying on the floor who was wrapped in heavy chains. The chains had been wound around him repeatedly in the manner of an old Charlie Chaplin movie. The man in the chains looked bruised and somewhat battered. Our conversation had awakened him out of a stupor and he moved a bit. I looked at him briefly and then realized that it was Mitch Cohen, my friend, crewmate and nose gunner on Con Job. I was startled, excited, comforted and relieved all at once.

"Mitch," I gasped, "it's you! Oh, am I glad to see you! Are you O.K."?

"Kovar, you're here," he mumbled in a labored and semi-coherent voice. "You're here. I'm glad."

The officer on duty interrupted our greetings and gruffly pointed for me to lie down on the floor and be quiet. Being in no position to argue, I stretched out under the desk and tried to relax. I was no longer in civilian hands and that was a relief. However, I knew that until one was in military hands and officially registered as a Prisoner of War, anything could happen. It was very easy for an individual to simply disappear if that person was not registered. With that said, to become an official Prisoner of War was no guarantee of safety but, it did set some restraints on what might happen.

I was now involved in some sort of a military system and Cohen and I were together. We were being accounted for and this was a comfort because it provided a measure of safety. We were a part of official machinery and someone was starting to keep track of us. Things were beginning to fall into place. I relaxed on the floor next to the desk and dissolved into sleep.

Chapter 5

A bustle of activity roused me into awareness. The day had begun in the garrison and people were going about their various duties. Men stomped in and out of the office involved in the scheduling and business of the day. My impression was that the soldiers in this garrison assisted in the work of the local farms. Part of their military duty involved feeding livestock, picking corn, and whatever other farm work was required. In the process of meeting the local needs and assignments of the day, Cohen and I were largely ignored.

I was stiff, sore, tired and hungry when I awakened from my few hours of sleep on the floor but Cohen was all of that and much more. He was bruised and aching from the beating he had taken when he had been captured. Further, he was cramped and miserable having spent the night wrapped in a heavy chain. His night had been one of pain and misery.

I turned to the officer in charge and, through words and gestures, I tried to encourage him to free Cohen from his chains. The officer was a middle-aged man who seemed quite sure of himself. He was very much in charge and apparently felt that the possibility of our trying to escape was near zero. Being in a confident and generous mood, he and his assistant unwound Cohen from his heavy chains. Cohen sat up dazed and subdued.

The whole community for miles around was well aware of the huge air battle and the bombings that had taken place a couple of days earlier. Everyone knew that many fliers, both Nazi and Allied, had been shot down and that it was necessary to account for all of them. Nazi fliers needed to be cared for and sent back to the job of protecting their country. Enemy fliers, like Cohen and me, had to be captured and punished. Subsequently, the officer led us outside to the courtyard where we could be displayed to the local townspeople. He probably wanted to assure them that the military was in charge and that escaped Allied airmen were promptly captured. The large, well-kept area was separated from the main street by a low stone wall. On our left was the building in which we had spent the night. Behind us were other old buildings which housed a kitchen and supply rooms. In front of us was a large pleasant lawn that extended to the low wall that bordered the main street.

Mitch and I could finally talk. We recounted our stories to each other. Although we had landed many miles apart we had been captured at about the same time. From this point on, our stories unfolded quite differently. In the preceding hours Cohen had been vilified and brutally mistreated after capture and he was afraid. He had been identified as both an American terror-flyer and a Jew.

Antisemitism was very real in this part of the world and hatred toward the Jews had been carefully cultivated for many years. Nazi propaganda, under the effective and demonic leadership of Joseph Goebbels, had attributed most of the world's injustices and hardships to the presence of Jews in society. The Nazis persistently portrayed Jews as evil beings who should be punished - even eradicated - and this idea had become deeply imbedded in the collective thinking of people in the Nazi world. By this time of the war, many Jews had been harassed or killed as a result of propaganda.

Because of his looks, Mitch Cohen became a recipient of this perverse and hostile thinking. He looked like the propaganda-based Nazi stereotype of a Jew with his dark curly hair, Mediterranean complexion and facial features. To make matters even worse, he was an American terror flier who dropped bombs on their country and sometimes, it was rumored, also dropped dolls and toys that were loaded with explosives in order to kill or maim children.

I was also hated for my crimes of dropping bombs but I was not a Jew. I had red hair, a fair complexion and a familiar Hungarian name. This put me in a different category, qualifying me to be a misguided human being who had somehow gone wrong.

News regarding the capture of American fliers had spread rapidly. We were put on display on the public green for the next few hours. Many townspeople took a walk to look us over. At times there were a dozen or more people standing outside the wall studying us. Most of the people were silent, although, a few cursed and shook their fists at us. Some people showed kindness toward us. One young girl, about fourteen years old with a bandaged foot, climbed over the stone wall and limped up to us. She smiled and gave us a few half spoiled figs and a couple of over-ripe pears to eat. This was quite an incredible thing for her to do, especially as we learned she had received her foot wound during one of our bombings.

Despite her kindness, Cohen and I continued to feel vulnerable and uncertain. We sat together in the warm sunlight of a beautiful morning, grateful to be able to lean on each other. The future was tenuous for both of us but especially for him.

A slender young man in his late twenties came up to talk to us. He was a Hungarian Jew and spoke excellent English. He stayed with us for at least a half-hour. During that time he spoke with us openly and frankly. He told us that, while our situation was dangerous, we would probably be turned over to some higher authority. He told us something about each of the townspeople who came by to stand at the stone wall: who they were and how dangerous they might be to us. He also explained that his job was to defuse unexploded bombs that had been dropped in the area. That was why he was being kept alive. Occasionally the fuse mechanism on bombs would fail to work and a bomb would not explode. Whenever this occurred, it was his job to defuse the bomb and keep it from being dangerous to the community. This was an extremely hazardous job and I was able to tell him how to defuse a particular bomb with which I was familiar.

I also told him that, since I was in Hungary, I might as well use my time constructively and try to learn the language. He responded emphatically by saying that the Hungarian language was almost impossible to learn unless one grew up with it. He said that if I wanted to study a language, German might be useful; anything except Hungarian. As for his own situation, he felt that since his job was a high priority, he had a pretty good chance of surviving the war even though the odds were against it.

Late in the morning a big, fat, jovial woman came out from the kitchen. She had one front tooth missing and was wearing a dirty apron. She carried a large pot of beef stew and a couple of bowls. She filled the bowls with stew and gave them to us. Cohen and I were both hungry and the stew was delicious so we ate our fill. The day was getting better and we began to feel more relaxed and secure.

Shortly after noon, two soldiers came out from the barracks with rifles hanging from their shoulders. One man looked strong, vibrant and in his early thirties. The other could have been his father and looked to be in his mid-fifties. The older man appeared to be a strong, gentle person, but he walked with a definite limp. He was obviously in charge and he pointed us toward the main street. It was clear that the two of them were going to guard and guide us to our next location. With the guards walking directly behind us, Cohen and I marched through town. Most of the townspeople stopped their activities and stared as we passed. A few of those who came out to see us seemed to be indifferent but most of them were hostile and some snarled words of anger. We were glad to have the protective presence of our guards. The older man had trouble walking and for the first hour we progressed fairly slowly. Then, as we entered another village,

Cohen and I agreed that we would pick up our pace, square our shoulders and walk proud and tall. We were American soldiers and we would march with strength and dignity.

Our guards kept up with us but the older man continued to have a bit of trouble. He appeared to have arthritic knees that pained him but he also had his pride and he tried to keep up. He did not order us to slow down: he simply gritted his teeth and kept hunkering on. After we had passed through the town however, the guards called for a stop. We sat down on the side of the road to rest while the older man recuperated. We continued on in that fashion for the next several hours.

In the late afternoon of this warm day, our guards decided to stop at a tavern in the village we were entering. All four of us sat together at a table in a comfortable bar. The guards ordered a tall glass of dark stout for each of us. We sat together in a relaxed and companionable mood and enjoyed the excellent stout. Our two guards chatted together while Cohen and I talked freely. Strangely, in the dark of this friendly place, we had the feeling of camaraderie. After having had a comfortable moment in the tavern, we resumed our roles as prisoners for the rest of the journey.

The long shadows of dusk were beginning to show when we walked into the city of Papa. The city appeared to be a large and important urban center. The tone of our relationship with the guards changed and became more military. Once again they became firm, distant, grim and in charge. Cohen and I were seen as dangerous enemies who should be hated and firmly dealt with. We marched a short distance through some sort of city park and finally arrived at a large two story building. The building was connected to a high concrete wall that covered a wide area. Obviously this was a major prison.

We entered an administration area and were directed up the stairs to the second floor. We walked into a room that looked and felt something like a county courthouse license bureau. We waited at the counter while our guards made their report and turned over our belongings to the woman on the other side. A moment later a Hungarian officer appeared and our guards left. The new guard was businesslike, efficient and hostile. He opened our packages and rummaged through them briefly. As he did, I asked for my New Testament which he refused; however, he did let me have a couple of my handkerchiefs and my dog tags.

A couple of hard looking soldiers arrived from the lower floor to take charge of us. Our new guards directed us down the stairs to the first floor and took us through several hallways and doors. Finally we entered into an open area that was obviously the prison exercise yard. It was large and bordered by a high stone wall about seventy-five yards away. We walked through the yard and left through a door on the other side. We entered a dim hallway and, after making a couple of turns, we finally stopped at a stout wooden door. One soldier coldly nodded for me to enter this cell, and I did. He closed the door and locked it firmly with a chain. Cohen and the guards proceeded down the hall where he was locked in another cell. The darkness felt like a blanket - an uncaring and unyielding heaviness.

My cell was black except for the suggestion of light coming from under the wooden door. I could tell the cell was small; the chill and dampness closed in on me. I smelled mold and urine but it also reeked of age, neglect and misery. I stood still in the gloom and tried to adjust to the forbidding atmosphere. Eventually I dragged my eyes from the thin strip of light seeping under the wooden door and my eyes slowly adjusted to the dark. Yes, the room was small and it contained an old, moldy straw sleeping bag. It was heaped on some planks which had been raised a few inches off of the stone floor. Near the bed, I found a bucket that contained a couple of inches of water which, I assumed, was for drinking. Next to the door was a battered bucket that served as a toilet. It was spattered with the ancient deposits of previous occupants.

It was quiet and eerie. There was nothing to distract me from my stark aloneness. A sense of powerlessness grew within me and I sat down with a thump on the damp straw tic. It smelled sour and I could tell that it was alive with vermin. I began to panic and realized that keeping control of my mental state was my new battleground. I tried to dampen my inner panic with the knowledge that my situation was temporary. I had not yet been interrogated by any real authority, I thought, and that was sure to happen. I lectured myself that my job now was to make-do as best I could and keep my spirits up. I tensely stretched out on the dirty ancient tic and made an effort to sleep. Each of the ensuing hours dragged by to make for a very long night.

Chapter 6

Like cold molasses, a bleak day slowly poured itself into being. I knew it was day because that hint of light under the heavy door was back. Occasionally I could hear the muffled sounds of movement coming from somewhere down the hall.

I had no way of knowing how long I would be here, but, it was clear that I would need some sort of discipline if I hoped to endure this place. Therefore, I reasoned, it would be a good idea to create some sort of activity schedule to occupy my time. I decided to do exercises, recite memory work and find ways of keeping my mind active and my spirits up. My problem was that I could recall very little in the way of memory work or poetry. I was hard-pressed to stay busy and I was lonely, worried and scared.

About mid-morning, I heard noises coming down the hallway toward my cell. There was the scrape and scratch of a key at the lock of my door and it was pulled open. Two glum, sullen Hungarian soldiers stood there. One of them handed me a bowl of non-descript soup and, without speaking, closed and locked the door. They fumbled on toward the next cell where I assumed Cohen was located. I guessed that this was the normal routine for this dismal place.

I came to treasure those breaks in the silence when muffled noises hinted at an outside world. That crack of light at the bottom of my door provided me a dim human connection. But these were small comforts and my efforts to stay busy were difficult and time slowly dragged itself along at snail speed. It wore me down, the never-ending day. But it was even worse when the muted crack of light finally faded away. I lay on my musty tic and tried to be brave.

Outside my dark door came a rustle of coming footsteps. This human sound focused my attention. Then came a key scraping in the lock and my heavy wooden door squeaked open. Two surly guards grimly walked in. It wasn't much of a connection, but at least they hadn't forgotten I was here. One watched me as the other reached for my empty bowl and, without speaking, exchanged it for another. Dinner had arrived. The door was closed and locked and the sound of the guards' footsteps slowly faded away.

For awhile, I had soup to keep me company. It tasted terrible and I allowed my mind to complain and brood as I drank the greasy lukewarm swill. It kept my attention off the utter darkness and depressed silence. I made the swill last as long as possible, but eventually another gloomy, foreboding night closed in on me. I lay on my damp and soggy mattress feeling abandoned.

The long sleepless night ended when the guards appeared with my morning bowl of something that passed for food. About mid-morning, I heard the outer doors opening along with the sound of brisk, energetic footsteps. A key rattled and scraped in the lock on my door and it squeaked open. A Hungarian soldier stood outside and studied me for a moment, then stepped into my cell and closed the door behind him. This was different.

The man was about thirty years old, a little bigger and heavier than me and he was relaxed and exuded an air of competence. He carried a rifle and stood close to the door. Using hand motions he indicated that he was peaceful and he motioned for me to sit down. He tried to talk to me in Hungarian but I couldn't understand what he was saying. It soon became evident that he was simply a soldier who was curious about the new American prisoners. He had decided to visit me to see what I looked like. His mood was friendly. We tried to converse, but, it was very difficult. We did not speak the same language and, in a real sense, we lived in different worlds. We were both interested in the politics of the time and we were both involved in the same war, but, beyond that, we had little in common.

During our groping to communicate, I finally understood him to tell me that President Roosevelt had started this war and that he was some kind of monster. His comments shocked and puzzled me. I countered his comments by explaining that it was Hitler who had started the war and not Roosevelt. We blundered around with our conflicting viewpoints for a time and got nowhere.

I admired the guard's rifle and indicated that I wanted to hold it. He smiled with amusement that I would ask him, an enemy soldier, to give me his weapon. He chuckled at this idea and shook his head in disbelief, wondering what kind of idiot I thought he was to give me his weapon. Nevertheless, the idea interested him and he paused for a moment. Then, feeling secure in himself, he decided to see how far I would go with my reckless idea. He made a clear and obvious show of removing the clip from his gun as well as the shell that was in the breech. Slowly and cautiously he handed me his empty rifle. Then he stepped back and warily waited to see what I was going to do.

I gently received his rifle and quietly examined it. Then, moving slowly and with deliberate care, I handed it back to him. He received his weapon back and smiled with amused satisfaction. Chuckling to himself, he backed out of the room, closed and locked the door and departed. Once again I was left in darkness but my spirits had been lifted by the interchange. I had made human contact and that made a world of difference.

The day languished on as did another long night. The next morning's usual breakfast was the same depressed routine. Around noon, I heard the clatter of footsteps and the now familiar scraping of a key in the lock of my cell. The door opened and a couple of Hungarian soldiers stood there. They looked relaxed and friendly and they beckoned for me to come out. I walked out of my cell and followed them down the hall to a door that opened to the outer courtyard. As the door opened the bright sunlight of a beautiful day splashed into the gloom. It took a moment for my eyes to adjust to the light, but shortly, I realized that it opened out into the exercise yard. I stood in the bright light for a moment and then heard a familiar voice speak to me. It was Cohen.

"Oh boy is it ever good to see you, Kovar," he said in a horse voice.

"It's great to see you too, Mitch," I responded with a relief that almost welled into tears.

After three days of isolation in a medieval dungeon, the joy and relief of hearing a friendly English speaking voice was beyond description. We were both almost overwhelmed at having a friend in this enemy camp. Eagerly and gratefully we talked and shared a profound camaraderie. We were two frightened and vulnerable people in a very dangerous and forlorn situation. We each drew a tenuous but vitally important support from the friendship and goodwill that we shared.

High prison walls surrounded the open courtyard. Guarding the top of these walls was a Y-shaped barbwire fence. Escape from this open yard would be extremely difficult - at best.

A group of about fifty Hungarian paratroopers were lounging about next to the prison buildings. Later we learned that they were on rest and recuperation after having been in combat on the Russian front for several months. These paratroopers had all, presumably, survived more than ten combat jumps. These were tough troops - as tough as they get. They had all been through a long tour of hell.

Apparently, a couple of these men had decided to bring Cohen out into the open just to see him. Once outside, Cohen had encouraged them to let me out into the open also, which is what they did. Several soldiers wandered over to look at us and then drifted back to join their comrades. Meanwhile, Cohen and I stood talking to one another with our faces toward the high concrete wall and our backs to the soldiers.

Suddenly a bulky Hungarian soldier came from behind, grabbed Cohen, and roughly pushed him about twenty yards toward the wall. The soldier then stopped short and spun Cohen around. In a voice that was bitter with hate, he snarled obscenities at Cohen. Radiating malice, he pulled his pistol out of its holster, pulled back the hammer and jammed the barrel into Cohen's ear. Cohen stood shocked, motionless and frozen with fear. A tense, electric moment followed in which time stood still.

The only thing moving was the blood slowly leaving Cohen's face. He was stiff with fear. Everyone in the yard waited breathlessly as the soldier stood with his cocked pistol jammed into Cohen's ear. Suddenly he dropped his arm and backed away from Cohen until he stood a few paces away from me. He glared at Cohen with hard eyes that blazed with hate. Purposefully lifting his pistol he deliberately aimed the sights at Cohen's head.

Cohen hadn't moved, hadn't breathed, was still and stiff. Now, in slow motion his knees gave out and he began to sag. Half-way to the ground, he caught himself. With defiant resolve he stood up straight, squared his shoulders, and locked eyes with the enraged soldier. The set of Cohen's jaw said spoke a universal language: "all right, you bastard, squeeze the trigger!"

A silent, brittle, tense eternity followed as the soldier stood with his pistol aimed between Cohen's eyes. Then slowly and reluctantly, the soldier lowered his pistol. A long pause followed. Then he let up on the hammer, jammed the pistol back into its holster, and abruptly turned to stride back toward his comrades. Our thundering hearts pounded in our ears. It was a long breathless space of time.

Then Cohen staggered toward me - and I toward him. We stood face-to-face, looking at each other stunned, aghast, and speechless. After a long, strained moment, one of the Hungarian soldiers came forward. Solemnly he escorted us back to our cells.

My heart was still hammering as I entered the bleak safety of my dark, dirty, but quiet cell. We were not yet under the protection of the Red Cross or the German government. Our situation was still uncertain.

Chapter 7

Although this night was also damp, cold and lonely, for a long time I actually valued my cell's safety. As I laid down on my dirty straw tic with its writhing life forms, I was overcome with thoughts of decay: decay within these walls, decay of the human spirit, and the decay of moral compasses. The forlorn drag of a desolate night settled in. Feeling cold, worried and depressed, fell into a fitful sleep.

A new day arrived at last. Once again the guards grumpily delivered a bowl of morning swill. Again I forced myself to be busy doing pushups and repeating a few verses of memory work, but, my heart was not in it. I felt as if I was accomplishing nothing and that weighed down my spirits. I sat. The darkness closed in around me.

About mid-morning, I heard the outer doors to the hall open and then the sound of a booming, cheerful American voice. The door to my cell rattled as it was being unlocked and then opened. A tall, lanky man with the big voice entered. He paused for a moment to let his eyes become accustomed to the dark as the door closed and locked behind him. I was grateful, relieved and happy to greet him. Here was another American airman. I was not alone anymore and the weight lifted a bit.

The airman's name was Jim Lantz. He was about six feet tall and appeared to be a year or two older than me. He was still wearing his high altitude sheepskin jacket. He was from Montana, a state that I would later call home for nearly fifteen years. Jim was a P-51 fighter pilot. His engine coolant had been shot out in a dogfight a few days earlier. He had managed to make a good dead-stick landing and set his plane on fire before escaping from the area. Jim had been captured a few hours earlier and was now being placed in this cell with me. We greeted each other with real warmth and camaraderie. The value of an American friend in a situation as bleak and foreboding as this dingy dungeon, behind enemy lines, is beyond the power of words to convey.

As we talked, it suddenly occurred to me that he might be a plant: a German spy who would try to obtain information from me. We had been warned that this kind of thing could occur and we had been instructed on how to conduct ourselves should we ever be captured by the enemy. It had been impressed upon us that any information that we gave to the enemy, however small or insignificant, could be of great importance to them. A small fact, if it is the right one, could fill out a whole picture for the enemy and greatly damage the Allied cause. Our instructions were firm: name, rank, serial number and nothing more.

At about the same time that I began to consider his true loyalties, it seemed Jim started to wonder the same thing about mine. Our conversation became measured and careful as we each tried to deduce whether or not the other was truly an American. We cautiously parried with one another until we each gradually decided to trust one another. Only then did we really relax and again let ourselves appreciate each other's company. Jim's presence in this lonely cell gave me strength.

It was early in the afternoon when we heard the noise and clatter of guards coming to open our cell, and Cohen's, to lead the three of us out of the prison. Waiting to meet us was one of the most nattily dressed junior officers I have ever seen. His knee high boots were highly polished and were possibly the most beautiful boots I had ever seen. Even the holster for his side arm was gleaming with polish and pride. Here was a man who was very conscious of his position and his appearance.

As the officer received us as his prisoners, he paused for a moment and carefully made a conspicuous show of drawing his pistol from its holster and examining it. He was equally dramatic about putting his gun back into the holster with the flap properly adjusted so that the pistol could be quickly withdrawn. Patting his pistol was his way of letting us know that he was a man of competence who would not put up with any foolishness on our part. He was proud of his assignment to escort three terror-flyers to the next stage of their imprisonment. He wanted us to know that he was a man who was equal to the job, in case we had any doubts.

The officer escorted us out of the building and pointed the way for us to go. The three of us walked side by side with our guard behind. We had not hiked far before we encountered civilians: men, women and occasionally a few children. The hatred that was projected toward all of us was very clear. Some people shook their fists or spit at us. But however much they hated Jim and me; they showed an extra measure of loathing toward Cohen.

It did not take long before the three of us agreed that it would be safer for all of us if Cohen walked in the middle, with Jim on one side and me on the other. This configuration put Cohen a bit farther out of the reach of the people who grimaced and shook their fists at us from the side of the road. This arrangement seemed to work somewhat better; nevertheless, as we walked down the road three abreast, followed by our elegant guard, we were the object of much hostility.

The ill will was very real and our guard began to lose his confidence. He worried that the malice and anger might spill over into some sort of an incident that he would have to quell. The officer's grandiose feelings, that he was a hero guarding three enemy prisoners, began to evaporate in the face of genuine confrontation. His insecurity was unsettling for all of us.

It took a couple of hours for us to reach the railroad station. It looked very much like stations found in small towns in the United States. Our guard escorted us into the waiting room and carefully positioned us in front of the large bay window that overlooked the railroad track. He ordered the crowd of civilians in the station to move back and create an open space. Jim, Cohen and I stood together at the center of the window area with our guard standing in the open space between the crowd and us. Gradually, civilians gathered around, in a fairly large half circle, to stare at us. The mood was sullen. When even more civilians arrived, a tone of smoldering hostility began to grow and become palpable.

In the midst of this hostility, an older man, about fifty years of age, stepped out from the crowd and spoke to me in excellent English.

"Where are you from in America," he asked?

I moved a few steps toward him. It seemed to me that he was an educated man who was interested in speaking English to an American and hearing something of an American point of view. I was interested in knowing where we were and what we could expect, and so, a conversation between us began.

He was a committed Nazi and imbued with the Nazi point of view. In the course of our conversation he asked if it was true that we American airmen were paid five hundred dollars each for every German plane we shot down? He also said that he thought it was reprehensible that we Allies tried to target children in our bombings. I, of course, denied both of these allegations and pointed out that this was propaganda. He, however, remained unconvinced.

As we talked, the crowd seemed to pause and relax as they stood quietly listening to our conversation. As this distraction continued, we began to relax and feel somewhat safer. I edged, step by step, closer to the man, in order to hear him better, as we continued to talk. This process continued for a time and soon I found myself standing several feet in front of our elegant guard and somewhat closer to the gathering crowd who had become engaged in our labored conversation.

Suddenly, the entire crowd of civilians in front of me tensed with shock. Something startling was occurring behind me. I spun around and looked back just in time to see an enraged German soldier rush out of the crowd and make a violent slash at Cohen with his bayonet knife. Cohen instinctively stepped aside from the attack and the soldier missed stabbing him by a hair. Our guard stood wide-eyed and stiff with fright for a moment. Then gingerly, and nervously, he stepped forward in an attempt to calm the enraged soldier. Eyes blazing with anger, the soldier reluctantly turned and, while muttering curses, made his way back toward the crowd. Cohen stood stunned and shaken by the incident, as indeed, we all were. With that, I suddenly became newly aware of our precarious situation. I turned and rejoined the company of my companions by the bay window. Our guard nervously paced back and forth in front of us. We all waited grimly and impatiently for the train to come.

Chapter 8

The passenger train arrived late in the afternoon. Our imperious, but, no longer sure of himself, guard directed us to go toward the rear of the train and board the last car. We walked down the crowded isle and found two empty bench seats that faced each other. The guard sat on one seat, while we three prisoners sat on the other facing him. Soon a soldier and a middle-aged woman took the seats next to him. The train, apparently, had now become completely full.

I am confident that almost everyone on the train knew that American prisoners were on board, but, no one seemed to care or pay much attention to us. The fierce malice that we had felt in the station was simply not present. The passengers on this crowded train were pre-occupied with their own affairs and were not interested in us. The mood of our situation had significantly changed. We were just three more passengers on a train crowded with people who were trying to get to somewhere else.

The motion of the train was smooth and occurred without the bumping and jolting typical of American trains. A soldier in the seat ahead of us was interested in chatting and he turned around to talk. He spoke very good English. The soldier had been educated in England and simply wanted to bone up on his English. During our conversation, he complained about the decline in food rations but, there was no doubt that his mind and heart were fully German. He had been a soldier on the Russian front and was now going home for a short leave before he returned to his unit in Russia. From his perspective, he considered the Allied air offensive to be cowardly. He reasoned that, because the Allies were now putting more planes into the air than were the Germans, the air fights were unequal. He had no explanation, however, for the German bombing of Warsaw where they had no opposition at all. I told him that I thought Hitler was a bad leader for the German people. He countered with his conviction that Roosevelt was a bad leader for the Americans. He reasoned that Roosevelt could not possibly be a great leader because, after all, "Roosevelt had never been a soldier."

In the course of our conversation, he assured us that we were fortunate to be in the hands of the German Government since they respected brave combatants and would be careful to treat us well in accord with the Geneva Convention. He was not sure where we might now be going but he thought that it could be to the city of Budapest which was a hub for many administrative activities. In any case, we would be treated well. It startled me to think that we might be interned in Budapest. This was the place of my father's birth and it was also the target on my first two combat missions. It would be ironic for me to end up there.[2]

Other people on the train came by to speak to us. All of them seemed to be friendly and merely interested in chatting with an enemy soldier. Among our interested visitors were some teen-aged girls. They gave us Hungarian cigarettes which were very strong but were much appreciated. One of the girls took a shine to Cohen and the two of them carried on a brief flirtation. Our guard was once again feeling confident and proud of his important job of guarding us. He stood up to stretch and, benevolently, allowed us to do the same. He even allowed us to walk down the center isle a short distance.

---

[2] Our first mission took place on July 27, 1944, which happened to be the birthday of my eldest sister. On that mission we were hit by flack as we entered the target area and had one engine shot out. During our second mission, we were again hit by flack and returned with two engines shot out. We considered Budapest to be a tough target and we had great respect for the military there.

I had a full bladder and wanted to know where I could go to relieve myself. I indicated this to our guard, who then led us out the rear door where we stood on the little platform at the back of the train. Our guard stood with us on the platform of the slow moving train: he was now feeling secure, self-important and in charge. Little did he know, that as we three prisoners stood alone with him on the back of the train, we openly discussed whether or not we should overpower him and jump off the train and seek to escape. We could have done this quite easily, however we were inhibited by the facts of life: we were deep behind enemy lines and not equipped to attempt a serious escape. We did not know where we were for sure, nor did we have maps or plans as to where we might try to go if we did escape. Further, we had no money or underground connections. These were the facts of our situation and, therefore, the odds of our escaping and then actually making it back to Allied territory were somewhere between extremely slim and zero. Moreover, if we were captured after an attempted escape, we could be certain to receive very rough treatment from which we would probably not survive. Therefore, our amiable conversation about overpowering our guard and attempting to escape ended with a decision to be compliant prisoners. Usually, the recognition of facts leads to a rational and right decision. This was one of them.

It was about one in the morning and dark when we arrived at the train station in Budapest. It was a large station where many train tracks converged. We got off the train at a point where dim lights provided just enough illumination to enable a person to move from one track to another. Our Allied bombing had made rubble out of most of the station and destroyed many of the tracks. Some of the tracks had been repaired, while many others remained a twisted and tangled mess. In general, the station was a complicated and disrupted system that somehow, almost miraculously, still functioned but with many delays and frustrations. The station was illuminated by dim lights that had been strategically placed here and there and the place was quite busy with people milling around in the dark, each trying to find his right place.

Our mood and that of the atmosphere had changed considerably. Everyone was tired, frustrated and irritated. We de-boarded the train under the scrutiny of our guard who was aware of the change in mood and was once again feeling nervous and uncertain. He wanted to keep us away from the townspeople as much as he could. He led us a short distance away from the boarding platform and into the middle of an area where several sets of tracks converged and which was illumined by a single forty watt bulb. There we awaited transport to our next location.

We stood alone among the tracks for a time but, soon, people on the boarding platform saw us and a few of them stepped toward us to get a better look. Slowly and quietly more individuals gathered until there were a couple dozen people gathered in a semi-circle. They were quiet but their mood was sullen and morose. Sometimes one would mutter a comment then scowl and shake his head or his fist. Gradually the group grew until, before long, a fairly large gathering of irritable, antagonistic people completely surrounded us. Their tone was one of smoldering ill will which gradually became more open and vocal. Our guard observed this growing mob with ever deepening apprehension and uncertainty. It became evident that the situation could easily escalate and get out of control.

Meanwhile, the three of us had noticed that, whenever one of us spoke loudly enough to be heard by the people, the crowd tended to pause and listen. This seemed to ease the tension. We began to speak more often and loud enough to be heard by the townspeople who were gathered around. Jim spoke a few words and the crowd stopped to listen. Cohen spoke and the crowd listened. I began to recite the poems that got me through high school: "The Shooting of Dan McGrew" and also "The Cremation of Sam McGee" by Robert W. Service, the Poet Laureate of Alaska.

> *A bunch of the boys were whooping it up*
> *In the Malamute Saloon.*
> *The kid that handled the music box*
> *Was hitting a rag-time tune.*
> *Back of the bar, in a solo-game,*
> *Sat Dangerous Dan McGrew*
> *Watching his luck, was his light of love,*
> *The lady that's known as Lu."*

I spoke loudly and dramatically and, as I did, the crowd paused and listened. They began to quiet down and the energy of malice started to diminish as they simply stood and listened. I repeated each of my poems several times and the gathering people became even quieter.

At long last, a squad of soldiers arrived to relieve our nervous guard and to take us into custody. Fully exhausted, with shoulders sagging from fatigue, our elegant guard gratefully released us to the soldiers. His assignment was complete and he left somewhat the worse for wear. Our new guards opened a corridor through the crowd and escorted us out of the dimly lit station and into the blacked out city of Budapest.

Chapter 9

The squad of Nazi soldiers escorted Cohen, Lantz and I through the dark of blacked out Budapest. Finally we marched our way to our destination, a large old stone building. We were directed inside and thumped through hallways and down a couple flights of well worn stone steps. We arrived at a dimly lit guard station where we were turned over to a big, middle aged soldier who was seated at an ancient desk.

The man appeared to be tired and in a sour mood. He looked us over for a moment, radiating a tone of contempt and smoldering anger. His English was limited but his message was clear: he considered us to be despicable creatures of the worst possible kind. He intimated that we deserved to be hung for our crimes. With gestures and utterances, he went on to convey the idea that this was probably what would happen at dawn. We were alarmed and hoped he was just venting his deep malice and hatred toward Americans in order to frighten us. This man was only a guard and was not likely to have any real authority. Further, we had not yet been fully interrogated and the military would certainly want to question us before making decisions regarding our future. We desperately hoped that this was the case, but, the fact was that we were still not officially accounted for either through registration with the International Red Cross or any military group of importance. Therefore, if someone chose to dispose of us, no questions would be asked. This man spoke with the conviction of one who would dearly love to do the hanging if he had the chance.

We were then led by the light of a burning torch through a side hallway to a wooden door where Cohen and I were pushed inside. The door was closed and locked and our jailers escorted Lantz to a separate cell before returning to their station. Never have I seen darkness so deep and impenetrable: there was absolutely no light at all. We groped our way around the stone cell to discover its size and contents. It was very small, about six feet by ten feet. It was cold and wet from condensation and smelled of urine, mold and age. Our cell was a true medieval dungeon. It was seeped in hopelessness.

We discovered a heavy plank positioned against the wall opposite the door. It was about a foot thick and three feet wide, slimy with mold and ancient growth. The plank was not level and was placed at an angle so that one could not comfortably use it as a bed or a chair. It was perfectly positioned for discomfort since it was high on one end and low on the other. Totally exhausted, we had no choice but to make do with what was there. Cohen and I crawled onto the plank and tried to find a way to lie down and sleep. Because of the awkward tilt and narrowness of the board, along with the dampness of the cell, sleep was not possible. We positioned ourselves as best we could and shivered in the dismal darkness, awake, cold and worried.

Several hours dragged by. I was silently fretting to myself in the utter blackness when Cohen spoke, "Kovar," he said in a horse whisper, "you're a Christian and I'm a Jew. Do you suppose that it matters"?

I brooded about his question for a time and then, in an uncertain voice, answered, "Mitch, you're a Jew and you will die a Jew. And I'm a Christian and I'll die a Christian. Maybe it's the same God."

He grunted in response and again the darkness wrapped us in its foreboding fingers. We shivered and waited anxiously for a new dawn and a new hope.

The sound of iron-cleated boots and the screech of rusty hinges announced the start of another day. A new beginning had finally arrived and we hoped that we would now find out what our future held. Stiff and apprehensive, we were escorted out of our bleak cell by a morose guard. He escorted us to the little office area, where the first guard had expressed his desire to hang us. Jim Lantz was already present and he greeted us with relief. We were each given a cup of ersatz coffee and a single piece of dry, black bread. Then we were ushered through several hallways, up some stairs, and out of the building into the mid-morning of a new day.

We were on a street in what appeared to be an old part of the city. Parked close by was a battered and dirty old bus. We boarded the bus and joined a group of about twenty-five or thirty American airmen. These men were also prisoners of war who had been collected from various places in the area. Some were still dressed in flight gear and all appeared to be new prisoners like Cohen and me.

As we boarded, I looked over the bus and was happy to see that Rick Turnbull, our pilot on Con Job, was present among them. Cohen and I gratefully and warmly greeted him.

Rick Turnbull, Pilot

Rick had been beaten during the process of capture: his face was bruised and he had a couple of loose teeth. He was feeling rocky and worse for wear but he was grateful to see us and to be with friends. We asked him what he knew about the fate of the rest of our crew but he was as ignorant about them as we were.

The driver of the bus was a small man who wore about a dozen wrist watches on each arm. Obviously, the watches had been taken from Americans who had come before us. The dash-board of the bus was piled high with several dozen sunglasses, cigarette lighters and miscellaneous items which had also been taken from previously downed airmen. The driver spoke English quite well and, with great enthusiasm, told us about a new weapon that the Axis powers were about to unveil and that it would change the air war in favor of the Nazis. The weapon, he explained, was a revolutionary new gun that would shoot a large net into the path of incoming Allied bomber planes. This net would entangle the planes and cause them to crash, making the Allied bombing ineffective. We considered this to be a ridiculous idea but he seemed to believe it and none of us choose to dispute him.

We drove through the city and up a hill that overlooked a good portion of Budapest. Finally, we arrived at a hospital where we unloaded the wounded in our group. As this was going on, some of the American men in the hospital looked out of their windows at us. Among those looking out of the windows we recognized Lynch, our waist gunner, and, Britton, our ball-turret gunner, both of whom had sustained severe burns as they escaped from Con Job. They would probably be scarred for life. They saw us and we thankfully waved in recognition of each other.

A doctor came out of the hospital, entered our bus and spoke to us briefly in excellent English. Apparently he was a humanitarian and wanted to know our names so that we could be reported to the International Red Cross. We responded with a sense of relief and the hope that our families would be notified that we were still alive and being held by the Germans. I later learned that some of these names, mine included, did get sent out by a ham radio process to various parts of the world. Also, a few weeks later through a very round-about process, my parents received notice of my capture and status as a Prisoner of War from both Europe and the Far East. They had received this news even before our U.S. War Department had officially notified them.

The bus resumed its course along the top of the hill. We were all now feeling much safer and almost buoyant. New beginnings were in process. Some of us felt that, as Americans, we should show some bravado and spirit. Several of us joined together and began to sing some of the marching songs we sang while in training. It was a good try but it didn't last very long: our hearts were not in it. Finally, the bus arrived at something that looked like a large castle. We later came to believe that it was some sort of penitentiary that was now being run by the Gestapo.

We were ordered off the bus and herded into the forbidding looking stone building. There we were gathered in a large room and ordered to stand at attention while several Gestapo guards thoroughly searched each of us. When it was my turn, I pulled my hands out of my pockets and held my arms up over my head. In my right hand I held my little pocket knife. The soldier examined my body and clothing very carefully. He looked in my shoes but did not think to look into my closed right hand in which I held my little pocket knife. We were then ordered to undress completely and stand in the nude while our clothes were being deloused. At about that time, I became afraid that my knife would be discovered and that a severe beating would follow. Fear began to well up within me as I debated whether or not I should continue to hide my knife in my hand or perhaps hold it in my mouth. However my courage evaporated and, at a discreet moment when the guard turned his head, I dropped my knife into the pile of clothes on the floor. In the months to come, I would greatly regret that moment in which I succumbed to fear and deprived myself of a knife; the most useful and necessary of all tools.

A knot of foreboding rose within me as we were led down the hall to be deloused and given a shower. I was filthy dirty and, no doubt, smelled badly. A shower would feel good. However, we were well aware of the fact that the Nazis had been using a shower room as a camouflage for a gas chamber and that many people had been exterminated this way. I did not really believe that this actually was the situation with us now since we were enemy combatants and not political escapees. Nevertheless, I anxiously held my breath until I saw fresh cold water begin to flow from the shower heads. Then, with a sigh of relief, I relaxed and relished and enjoyed the luxury of a shower.

Following this wonderful treat, we retrieved our clothing which had been deloused with some sort of powder. Then, half-dressed, we were led to our separate cells on the third floor. At this point Mitch Cohen and I became separated. We would not meet again for about forty years when we made contact and resumed our friendship. He ended up in a Prisoner of War camp for enlisted men in northern Germany near the North Sea. During the last few months of the war, he and his companions marched south ahead of the Russian advance and traversed nearly the entire length of Germany.

Carrying my shoes and shirt, I was led up two flights of stairs to the third floor and marched down the hall to cell number thirty-three. The heavy wooden door with its thick iron straps was opened and a grim soldier, with narrow eyes and a brusque attitude, nodded for me to enter. I entered the small empty cell, the door slammed shut behind me and the lock was banged into place. The metal-cleated boots of the guard clomped away into the distance and gradually the sound faded into emptiness and silence.

Chapter 10

My eight by ten foot cell felt small and confining. The old gray concrete walls still carried a few flakes of white wash from some earlier generation. The cell was dirty and told a story of neglect and isolation. A single small window, near the ceiling, cast a diffused and uncertain light across the room. I sat down on the matted old straw tic that covered a narrow cot placed against the wall. In a dejected mood, I studied the thin, filthy blanket that was heaped on the end of the cot. It showed evidence of lice and other vermin living an active life there. The room was empty, except for two buckets placed near the door: one bucket held drinking water, while the other was for refuse. I was in another dungeon - another cell that reeked of despair and hopelessness.

I was in a large prison and part of some sort of official system, having had my name registered with the International Red Cross. Being part of that system was comforting to me, in that I was probably now much safer. I had no idea as to what might be ahead. Whatever the future, however, I reasoned that it was important that I keep myself in as good shape as possible. My best defense for the interrogation and the uncertainties that were yet ahead would be to stay in good shape.

Late in the afternoon of this uneasy day, I heard the click of hobnailed boots and the rumble of a cart moving along in the hall. The noise paused at the door to my cell and a morose soldier opened the wooden, shoulder high peek-hole within the heavy main door. He grunted something as he pushed a dirty aluminum bowl through the opening. The bowl contained an insipid soup that appeared to have been made by crumbling a few loaves of heavy black bread into a tub of hot water and then adding a couple of pounds of lard. It was nearly tasteless but it was food, and I ate, or rather, drank it. Then I again sat on my bunk in silence. It was difficult to keep despair from taking over as the sun slowly ebbed away and died. Darkness fell. I wrapped myself in the dirty, sour smelling blanket and sought solace in sleep.

I awakened with the dim light of dawn coming in through the small, high window in the outer wall. I set about the job of trying to be busy and active: I did pushups, calisthenics and repeated the limited memory work that I knew. I counted the cracks in the floor and walls, studied the lines of lice larva that laced my grimy blanket. If a bug skittered across the floor, I would tease and herd it; anything to keep active.

About mid-morning the soldier with the hobnailed boots and sour disposition opened the serving window again. He received my empty metal bowl and slid another through the opening. This new bowl contained a concoction that was as equally nondescript as the soup served the night before. I ate it and then continued in my efforts to stay busy. The place was eerie and quiet until late in the afternoon when the evening meal arrived. This lonely schedule continued for the next four days. It was hard work trying to avoid the ever present gnawing of fear and despair that sought to flow in and take hold.

Late in the morning on the fifth day, I heard the clicking of hobnailed boots clomping down the hall. I was busy at the task of keeping busy. I was doing pushups and my shirt was lying on the cot, when suddenly I knew that the footsteps would stop at my door. I jumped up and hastened to put on my shirt and groom myself. I tried to look sharp and military. My shirt was half on when the guard unlocked the door, opened it and beckoned for me to come out. I walked out of my cell, still trying to tuck my shirt into my trousers.

As I worked at trying to look confident and military, the guard marched behind me. We went down the hall and around a couple of corners until we arrived at an office area. The guard on duty there looked me over, made a quick inspection for weapons and then nodded for me to enter into an inner office. It was large, spacious, well lighted and carefully decorated. The impeccably dressed Nazi SS officer sitting behind a desk stood up respectfully as I entered. He was a strong and competent looking man of about forty years of age. His large mahogany desk was heaped with American cigarettes, watches, G.I. chocolate bars and various other items that had been taken from prisoners before me. I stood at attention and saluted smartly. The Nazi officer casually returned my salute and sat down. Then he said, in flawless English and with the easy confidence of one who was fully in charge, "Good morning. I hope our hospitality has been adequate for you. I also hope that I will not have to detain you very long. By the way, if you wish to smoke, please help yourself," and he pointed to the table.

I nodded to him and looked at the table. It was piled high with cigarettes, cigarette lighters and miscellaneous items. I eagerly took a cigarette from the pile on the table and lit it with one of the matches present. He patiently waited for me as I took a couple of deep drags from my cigarette. Then, sitting back in a relaxed mode, he said, "I have to ask you a few questions and I hope that you will cooperate. I don't wish to detain you too long because I am confident that you are an American airman. Nevertheless, I have to question you anyway. I must confirm that you are not a political escapee trying to avoid capture. We must be sure that this is not the case. Tell me your name, rank and serial number."

"Sir," I responded, "I am 2nd Lieutenant Leonard J. Kovar," and I gave him my serial number.

"Thank you. What is your group and squadron number"?

"I'm sorry sir, I can't answer that."

He paused patiently and then said, "You should answer me because I know the answer anyway. What were the tail markings on your aircraft?" he went on.

"I'm sorry sir, I can't answer that," I replied.

"It would be better if you were cooperative and told me because I know anyway. Who is your commanding officer"?

"I'm sorry sir," I said, "but I can't answer that."

"What was your target on the day you were shot down"?

"I'm sorry sir, I can't tell you that."

My interrogator was beginning to look firm and somewhat irritated. "Why don't you tell me?" he asked. "I know all of this anyway. I am only trying to be sure that you are not a political prisoner seeking this method of escape."

"All right sir," I responded, "if you know all about me, then you tell me."

He looked surprised and reflective. He paused and after a moment of hesitation said, "All right, I will!" He reached under his desk, fumbled around for a bit, and then pulled out a very large, thick ledger book and placed it on his desk. He opened it and carefully paged through until he found the place that he wanted. He pressed his finger into the book and began to speak.

"You are 2nd Lieutenant Leonard J. Kovar. You were a part of the 451st Bomb Group and the 727th Squadron. You joined your aircrew in Salt Lake City and trained at Tucson, Arizona. You were commissioned on January 15, 1944. Your last mission was your tenth. You were born October 24, 1922, in Minneapolis, Minnesota in the Swedish Hospital. You went to Roosevelt High School..." and on he went for a couple of minutes stating details about my life that I had mostly forgotten. He rattled off all of the names of my crewmembers except for the photographer who had joined us just moments before takeoff. He turned over a few pages and read the names of the chain of command for our squadron and our group. In all details he was completely correct, except that his listing of our chain of command was a few days late.

I was astonished and, with a voice stammering with incredulity, I asked, "Where did you get this information"?

"Oh you Americans," he said with a tone of superiority and contempt, "you're very stupid: you print all of this. We merely compile it."

He called for the guard to come and take me to the transport cell. He looked at me and said, "You will go to a camp for airmen. For you, the war is over."

I felt relief as the guard marched me down the hall and around the corner to the officers transport cell. I now had the security of being formally in the system as a Prisoner of War. We arrived at the proper cell and the guard opened the door. I entered and the guard locked the door behind me.

The small room was occupied by about a dozen American airmen. Among them were my crewmembers: lead pilot Turnbull, second pilot Warren, and navigator, Gould. We greeted each other with great relief and camaraderie. Everyone talked at once as we told of our experiences and tried to catch up on what we knew about the other crewmembers. Between us we managed to account for everyone except our engineer and top-turret gunner, Jack Roach.

Turnbull had serious doubts that Roach ever got out of the airplane. Turnbull had glanced at the top turret as he rushed to bail out of Con Job and saw Roach slumped over his gun with blood coming out of his mouth. Roach was never heard from again and was later listed as killed in action.

Bette and Jack Roach (married two days), he was killed in action during the mission in which I was shot down

About dark, I heard the staccato noise of military boots coming down the hall. The key rattled in the lock and two armed guards opened the door and motioned for us to come out of our cell. They directed us down the hall to a different cell where a number of other American airmen occupied the small room that measured about twelve feet by twelve. We entered the crowded room and were greeted by groans from the men within. They did not want us to join them. There were already eighteen men in the room and now, with us, there were twenty-one men jammed together. The room contained five blankets, six straw mattresses, one bucket for water, one bucket for refuse and one tiny window high up on the wall. It was a small cell for all of us to occupy and we were extremely crowded.

Somehow we absolutely had to get along in this confined and miserable situation and to accommodate as best we could. In the next hour or two, after much exasperating trial and irritation, we found that we could all sit down at the same time with reasonable discomfort. Hours later, after more trial and error, we discovered that, if we positioned ourselves properly, we could all lie down at the same time. It was a tight squeeze and we were uncomfortable, hungry, tired and afraid. Tempers grew short but we all realized it was utterly essential that we try to cooperate. It was a grim and anguished setting that was very close to being intolerable but, it simply had to be endured.

Late in the day, we were each given a couple of slices of some very good black bread. We understood that this bread contained about one-fourth sawdust, along with a mixture of wheat, oats and rye. Presumably it was months old and had been baked in a quick, hot oven to seal the crust outside while the inside was still somewhat raw. Baked in this way, the bread stayed damp and spongy inside but, strangely enough, did not mould. It was a basic food for the German army and was quite nutritious.

Several men were added to, or, subtracted from our group during the next five days. Our stay in the transport cell was an experience in misery that sometimes bordered on anguish. Fleas and body lice were a constant presence and we were far too crowded for any measure of comfort. Moreover, our twice daily diet of bread and lard soup, served in filthy bowls, was unappetizing to our still delicate taste. Almost everyone had some measure of dysentery and a couple of the men had pneumonia. The guards brought in a handful of charcoal pills for use as medication for those who were the sickest. One of the men in our group had a New Testament and it was in constant use. Another man came up with a set of chessmen molded from doughy bread pieces.

Our circumstances were miserable, and looking back on this, I wonder that we endured it at all. We were crowded, underfed and half sick; nevertheless, our morale held up surprisingly well. We were grateful for having gotten this far and we were now being processed through an orderly military system. Our present situation was temporary and we each felt that we had beaten the odds and were lucky to be alive. We steeled ourselves to endure. This experience is a good example of the fact that a person can endure almost anything, if he makes up his mind to do so and if he believes that there is an end in sight.

At about ten o'clock on the fifth night, I heard the clomping of a squad of guards coming down the hall. The lock on the door scraped and jangled and the door swung open. We were ordered out of our cell.

The really sick men were separated out and sent to an infirmary, while the eighteen or twenty of us left, were marched down the hall until we reached the outside of the prison. We were crowded onto a truck and driven through Budapest to the railway station where we were herded onto a forty and eight boxcar[3].

We prisoners crowded into half of the small boxcar while our three guards stretched out in the front half of the car. Several bails of straw were spread out for our guards to use as bedding while we prisoners were given one bail of straw to share among us. We were also given one long link of baloney, a couple loaves of black bread and four Red Cross parcels.

Somehow it fell to me to try to distribute these items. This was a difficult task and it involved bickering and compromise. The job finally got done but no one was satisfied. The Red Cross parcels contained real food which we thought was delicious. We shared it as best we could but we did not yet know how to use it to its greatest advantage. That would come later.

---

[3] The term forty and eight boxcar was the name the doughboys of World War I gave to the boxcars. They are considerably smaller than our American boxcars and it had been determined that each one could hold a maximum of forty men or eight horses.

Red Cross Package*

There were numerous stops during the confining, hungry and uncomfortable four days that we were in the boxcar. During this time, the guards were changed several times and occasionally, the boxcar would be sidetracked for awhile. On one of those stops, a guard went out and returned an hour later with a large bucket of boiled beans. He had brought the beans back as food for us prisoners to share. We only had four aluminum cups available among us and we tried to dole out this food equally. Somehow, it again fell to me to ladle out this wonderful delicacy. I did so very carefully; every eye watching to make sure that no one was shorted and that everyone shared equally. At another stop, one of the guards came back with two pails of beer which I also ladled out with great care. We savored the precious liquid to the last drop.

The hours and days passed for us much in this fashion. Then one night, when our boxcar was sitting on a siding at a railhead in Vienna, air raid sirens suddenly split the air with their mournful wails. Moments later, German arc lights began to search the sky for enemy bombers: a major British night bombing attack had begun. Immediately we all became uneasy because this very railhead could be one of the British targets. The British were not very accurate with their bombing, especially at night, but occasionally, they did hit what they aimed at and we were in a sensitive area. Soon I could hear bombs explode in the area. I sat tense and nervous in our boxcar and listened to the rumble of explosions that occurred around us. None, however, exploded in our immediate vicinity.

Late that night, when things quieted down, our train continued its erratic jostling toward a destination that was unknown to us. My attempt to sleep on the hard floor of the shifting boxcar was fitful at best. About three in the morning I awoke from a short nap and sat up against the back of the boxcar. Another man also awoke and sat next to me. We began a conversation that became a decisive moment for my life. We were talking about American automobiles and which company made the best car. As we chatted, suddenly out of nowhere, a thought flashed into my mind that had nothing to do with our conversation or our immediate situation: by golly, if I survive this war, I am going to go to college and I will graduate. This was a very important cross-road event in my life. How and why this thought erupted into my consciousness at this time is a mystery to me, but, it shaped my destiny.

It was early evening on the night of September 6, 1944 when the train stopped on a siding near the town of Sagan, Germany. This was the location of a major prisoner of war (POW) camp for American and British airmen. I felt both fear and relief as we were ordered to de-board the train and line up to be counted. A couple of squads of grim-faced German soldiers with guard dogs lined the track to see that no one got out of line or tried to escape. The tone of these front-line combat soldiers, who had just returned from the Russian front, was businesslike and no-nonsense. It felt far more menacing than the fairly relaxed attitude of the guards who had been with us on the train. These soldiers sent a message that was loud and clear: no foolishness or deviation from orders will be tolerated. Anyone who gets out of line, even just a little, will be shot. Period!

A German officer stood on a little platform and lectured to us in excellent English. "Gentlemen, you are now prisoners at this elite camp, Stalag Luft 3. We expect you to behave like gentlemen. Any disruption by you will be dealt with immediately and harshly. You should know from the start that the punishment for an attempted escape from this camp is severe, quick and certain. There will be no exceptions. Our policy in this regard has been determined by the _British Commando Handbook_. This manual states that an Allied soldier on German soil is expected to live off the land. Living off the land would involve stealing German goods. This, obviously, is sabotage and the punishment for sabotage is death."

He amplified this, as he continued to speak in this vein, until he felt that he had made his point. We prisoners accepted his lecture without surprise. After all, we were in a POW camp, not a day camp for cultured gentlemen. He did surprise us, however, by saying that a few of us would be permitted to send a cable to our families if we wished. A short time later, Gould, Warren and I sent a combined cable that we were okay and were now prisoners held by the German government. This message got through to our families in December. It was delivered over a ham radio from somewhere in Europe and also the Far East.

After we had been counted, we were lined up three abreast, with guards flanking both sides of us. We marched into the camp through a double set of barbwire gates. We were then directed to the Cooler for our first night. The Cooler was a separate jail within the larger camp for prisoners who had gotten out of line and needed cooling off before admittance back into the main camp. We were locked, two to a cell, for the night. As I curled up in the corner of my cell, my feeling was one of relief and gratitude: relief from immediate danger and gratitude for a sense of safety. My companion and I each stretched out on the floor to sleep.

Stalag Luft 3 POW camp near Sagen, Germany [4]

---

[4] All of the pictures in this book that are marked by a * were taken by a fellow prisoner of war with a hidden camera.

Chapter 11

Stalag Luft 3 was well designed and orderly. It included certain amenities, such as, a small library and a hall for lectures and movies. The camp, however, had decayed a good deal during the years of war. By the time I arrived it was overcrowded, drab and radiated a spirit of subdued anger and resentment; nevertheless, it was laid out in an orderly way. It was made up of five separate sections, each surrounded by an elaborate webbing of barbwire fences and guard towers. Each section held about two thousand American or British airmen. Outside of the compound were the German administrative and warehouse buildings. Beyond the intricate webbing of barbwire fences circling the entire camp, a dense forest spread in all directions. The Oder River was close as was the town of Sagan a few miles away.

It was mid morning when a guard opened the door of my cell and, in a businesslike way, motioned for me and my cellmate to come out and line up with the other new prisoners. I felt a sense of relief in now being in an official place with an orderly system. We were ushered outside and directed to a processing building. As I stepped outside and looked beyond the high barbwire fences, I noticed that the woods looked clean and peaceful. They were a stark contrast to the drab and depressed looking barracks of the main camp that I would soon enter.

Upon entering the processing building, we were ordered to undress and toss all of our clothing into a pile. We stood naked awaiting our turn to enter a small room to be dusted with a delousing powder. Then, along with the other new prisoners, I scurried through a supply room and was given an aluminum bowl, a table knife and a spoon. At the next room, clothing was tossed to me that had been taken from some earlier prisoner who had passed before us. The process was simple; as we passed by a long table, a German soldier glanced at each man and then tossed out clothing that he thought might fit. The process was fast and efficient, but no one received clothing that fit very well.

I was reasonably satisfied with the clothing that had been tossed to me, except for the shoes. They were old, worn and poorly made. More importantly, they did not fit and were about two sizes too small. This concerned me a good deal. Ill-fitting shoes would be uncomfortable, cause blisters and be cold: they would limit my ability to walk. I knew that the single most important piece of clothing or equipment that I could have in this situation would be good shoes. If my feet did not function I would not get very far, and this worried me.

We dressed in our newly issued clothing and gathered together in an adjacent room. About an hour later, a neatly dressed German officer entered and took charge. He ordered us to be seated on the floor. Then, standing before us, he presented an orientation lecture regarding our place in this camp. Speaking excellent English and in a calm and almost benevolent tone, he told us that we were now in our permanent camp. He said that for us the war was over - provided that we were sensible and did nothing foolish. He pointed out that we should be thankful that we were safe in this place and no longer involved in the fighting. He assured us that the German government wanted us to be well and comfortable. He stated that everything possible was being done to keep us healthy and safe. Stalag Luft 3, he emphasized, was an elite camp where we would find many opportunities for study and contentment, as long as we were cooperative.

He advised us to take advantage of the good things available at this place and not cause trouble. He pointed out that this camp was for gentlemen air officers and was a comfortable place to be, and that it provided opportunities for gracious living. For example, he pointed out, it had a library where we could study and prepare ourselves for a profession after the war. He further noted that there was a theater for cultural activities, as well as, a parade ground that allowed for exercise and sports. He went on to point out that; of course, the camp had rules and schedules which must be maintained. He assured us that if we cooperated we could have a comfortable, quiet life and that for us, the war was over.

His tone changed to one of sorrow as he went on to say that occasionally some men were unappreciative, foolish and attempted to do radical things, such as escape. This, he pointed out, was completely unacceptable and would be the worst thing that anyone could do. For starters, he declared that escape was simply not possible: the camp was too carefully planned and guarded. He noted that every precaution had been taken to make escape impossible. Furthermore, he added, that any attempt at escape would be very dangerous to our fellow Americans. It would compel the Germans to act in an extreme way and force them to make reprisals on innocent prisoners.

In a fatherly tone he went on to say that all of this was in accord with Allied standards. Then he, too, read that section from the *British Handbook for Commandos* which stated that if an Allied soldier found himself behind German lines, he should sustain himself by living off the land. Just like the previous guard, he reasoned that living off of the land would involve stealing or using government equipment. This was clearly an act of sabotage and sabotage was, of course, punishable by death. "Therefore," he concluded, "don't even think about escape, because if you try, you will be shot at once with no questions asked, and others might have to be punished as well."

He continued in this vain and urged us to be rational, patient and cooperative. He assured us that if we did this, life would then go well for everyone.

By the time he concluded his lecture we thought he had made a good impression and had presented a fine speech. I felt relief and gratitude that I was officially registered and safe in a major prison camp for Americans. It was true that a person would not likely be shot or turn up missing without notice at an organized place like this. He was sensible and very reassuring. Having finished his lecture, he turned us over to a sergeant. A squad of soldiers was to oversee our entrance through the barbwire gates into our home for the remainder of the war: West Camp of Stalag Luft 3. They escorted us through the double set of barbwire gates that entered the West Camp. Once we were inside, they closed the barbwire gates and left, leaving us standing as a group wondering what to do.

Just then, an experienced kriege marched up and took charge of us. 'Kriege' was the term used by the prisoners in the camps for one another. He referred to himself as Adobe Dick. He was a big man, about thirty two years old, who flourished a long handlebar moustache that had been carefully shaped over many months. He oozed the confidence of a tough, angry person who had plenty of authority and would not hesitate to use it. He briskly ordered us to follow him. After leading us into an open room, he irritably ordered us to be seated on the floor. We did this and sat quietly while this angry kriege paced back and forth in front of us and muttered curses under his breath.

We new krieges sat quietly in front of him. He stopped his pacing, stood and glared at us angrily. Then speaking in a harsh tone, he informed us that he was a Major and that he outranked all of us. He stated that he was a part of the American chain of command within the camp. He informed us that he was an aide to the senior American officer, Colonel Iron-Ass Alkire, and that Colonel Alkire was the person from whom we took orders. He paused again and then, pacing up and down for a moment, he glared at us and profaned in a firm and bitter tone, something as follows:

Now you assholes have just been captured by the Germans. You were shot down or you crash landed and I don't need any more details because I've heard it all before. So you damn near got killed two or three times and you were scared to death. Now you are here in this nice, safe, cozy camp and you dumb bastards are just idiot enough to feel grateful to these god-dammed goons.

You have just had a lecture on how lucky you are to be here and alive, and right now you are feeling sooo grateful and sooo cooperative to these god-damned shitheads who, just yesterday, were busting their gut trying to kill you. Now get over that damned stupid feeling of being grateful or thankful to these vicious bastards who started this war and who are not done working you over yet. Now get this! You are American soldiers: you are under our chain of command and these god-damned German murderers are the enemy. Don't you ever think otherwise. They would kill you in the blink of an eye if they had an excuse and never forget that. They are the enemy!

"A couple months ago," he went on, "after an escape by men in the British compound, these damn goons lined up fifty of our men and," pointing angrily towards the woods outside, "machine gunned them down. Killed them all. These damn goon bastards will do the same to you if they find an excuse."

His forceful monologue continued for about half an hour until he felt that he had made his point. He had and it was very sobering. The fact was, we were American soldiers in the middle of a war. Our job, as American soldiers, in or out of a POW camp, was to disrupt the enemy in any way possible. We were still American military men at war and Germany was the enemy.

His lecture had brought me and my companions up short and it re-arranged our attitude. Whatever goodwill and gratitude I was beginning to feel over the preceding few hours quickly dissipated. He was absolutely right. We did not owe the Germans one whit of gratitude or good will. They were the enemy. We were militia in the middle of a war and the war was not over for us or anyone else. Stalag Luft 3 was not a rest camp for gentlemen officers. It was a grim prisoner of war camp and our stay here was an interval in a dangerous story that was still unfolding.

Still speaking briskly, Adobe Dick went on to tell us that the camp was overcrowded and that all of the bed space was occupied and that new bunks were being built to accommodate us. Meanwhile we would be sleeping in tents that had been placed between the buildings. Then he ordered us to follow him across the compound to the cook shack to have our first meal.

A wide angle photo of Stalag Luft 3 near the town of Sagan and the Oder River*

*Stalag Luft 3 near Sagan, Germany**

*Winter in Stalag Luft 3**

Stalag Luft 3 and the forest just beyond *

A very cold winter at Stalag Luft 3 *

This picture looks almost idyllic, a beautiful moment on film contrasted with the reality of being a prisoner of war on a bitterly cold day.

Only those with a job to do were outside. The rest of us were huddled beyond these thin walls trying to stay out of the wind.

Chapter 12

I lined up with the other new prisoners, now called krieges, outside the cook shack located at the center of camp. We were about to receive our first meal, one very thin quarter-inch thick slice of heavy German bread with a suggestion of ersatz jelly spread across the center. All went well until the last man stepped up to the window to receive his ration of food. There was none left for him. The kitchen was short one slice of bread. Suddenly a big argument erupted in the cook shack. How could they possibly be short one piece of bread? What had happened?

This big flap continued for quite a while until finally, at long last, a thin slice of heavy bread with a slim layer of ersatz jelly was produced for the last man. Thus began my introduction to the fact that food was a high priority in this place and that every morsel was always accounted for.

West Camp was made up entirely of American men, a few of whom had been prisoners for two or three years. My group was one of the last groups of prisoners to be assigned here, and crowded or not, room had to be found for us. The krieges who had been here as POWs for two or three years were not pleased to see us since our arrival crowded and complicated things. More importantly, our presence cut into the limited food supply. I slept in a tent set up between the buildings during the two or three weeks that the barracks were changed from double to triple bunks. I was assigned to Room 13 in one of the barracks and I took my meals with the men from that room.

We were fed one main meal per day. Each meal was prepared in the central cook shack. The meal usually consisted of a thin warm soup and sometimes a piece of potato. We made a separate snack mid-mornings and late afternoons from our German bread ration of thirteen men per loaf, per day. The loaf of bread was about the size of an American one pound loaf, but it was much heavier and weighed about three pounds. Presumably it was made out of equal portions of wheat, oats, bran and sawdust. It was actually rather good bread and was quite nutritious. Along with the thin slice of bread we would sometimes have a small ration from a Red Cross parcel. During the early years of the war each man had received one Red Cross parcel per week; however, this allotment had been reduced to one parcel per four men per week. Overall the food ration was minimal and everyone lost a good deal of weight. I was always hungry.

I recall my first meal with my group in Room 13. The fifteen of us sat on a bench around a long table and, in the mid-afternoon, two men were sent to the cook shack to get the bucket of soup and potatoes for the men in our room. When the food was brought into our room it was carefully ladled out so that each man received exactly the same amount. This same scene would play out day after day; however, occasionally there would be small pieces of meat or gristle in the soup. In that case the pieces were carefully distributed. If there was an extra piece left over it would be set aside and we cut cards to see who would get it.

At my first meal, one of the older krieges who was called The Chief, because he had a bit of American Indian ancestry in him, asked me to bring the group of men up to date on how the war was progressing. I naively assumed that these men, having been isolated from the outside, had no information about the progress of the war and I would, therefore, bring them up to date regarding the state of events. With a certain sense of importance and pleasure I proceeded to update the group about the war's progress. I explained how the Nazis were in the process of withdrawal and consolidation on all fronts and how we could expect victory in three or four months. I was very eloquent, knowledgeable and convincing; I thought. A couple of further questions were raised by some of the men present and I obliged with the answers;

however, something did not seem quite right. Nevertheless I continued to confidently answer questions. Occasionally The Chief or one or two others seemed to suppress a snicker and this puzzled me. Later that afternoon when the daily news update was passed through the camp, I was surprised to find that the Germans supplied the official war news every day. More than this, our American X Organization within the camp, through its clandestine radio, regularly picked up the American and British reports as well. In other words, these krieges were far more informed than I was.

Knowing that they were all already well informed, The Chief had set me up for ridicule and to make me look like a fool. He knew that my knowledge was incomplete, as well as, being a few weeks late and that the men present had a far better picture of what was going on than I did. This deception to set me up for ridicule hit me hard. In the weeks ahead, I began to feel insecure about how to fit into this place with its established organization and cliques. To be set up and made a fool puzzled and shocked me. I had taken it for granted that the Germans might threaten, mistreat and even kill me - but I did not expect to be mocked and made to be a fool by my fellow Americans. However, this is what had occurred and it shook my confidence and sense of belonging.

The fact that I was surprised and stunned by The Chief and his undercover mockery says more about my youth and naivety than anything else. Put any group of men in a stressful situation and keep them uncertain about the future, let them be hungry, cold and deprived of meaningful activity and you have the perfect setting for anger, resentment, cynicism and depression. These feelings always find ways of projecting and expressing themselves in the process of living. Nearly always this expression is negative and comes out as sadism, aggression and the like. There was plenty of this present at Stalag Luft 3.

I had simply become the target of The Chief's negative energy and that of a couple of other men for a time. In my naivety, I reacted to this much more personally than I should have. It established a feeling of insecurity within me that greatly shook my self-confidence during my three month stay at Stalag Luft 3. For this reason West Camp, for all of its stability, was more unsettling than some of the truly dangerous and unpleasant situations that I had experienced and which were yet to come. My experience was shared by others, of course, and there were some men who became emotionally spent and went around the bend, as we called it. The psychological stress was difficult for everyone.

I moved into Room 13 when the triple-deck bunks were completed and fell into the routine of the camp. At least twice each day we assembled in formation for a body count. This was called 'Appel' and it took place on the playing field. We prisoners lined up in a prescribed fashion. Then the German guards counted and double counted to make sure that no one was absent. While this was going on, a special squad of guards ransacked and inspected various rooms where they looked for signs of escape activities. We often stood in the field for an extra hour or two while the guards completed their careful room inspections.

As fall gradually merged into winter and the weather became colder, the Appels seemed to become longer and occurred more frequently. I recall standing in formation with my feet cramped and hurting in my tight and ill-fitting shoes while I shivered in the cold. I looked enviously at the guards who were dressed in their heavy winter clothing. I particularly envied the German Commandant who would appear wearing a beautiful ankle-length leather overcoat with a warm comfortable fur collar. I soon developed what we called chill blains on both feet because of my poor shoes. The chill blains looked something like blood blisters or warts that became hard black spots where frozen flesh had died. They were not painful, but they certainly were not normal. I had never seen them before nor have I seen them since. I wondered why they did not develop into gangrene. Fortunately they did not.

Two interests consumed all of us night and day: food and the progress of the war. We talked about the progress of the war endlessly and in detail. We also spent a good deal of time jotting down lists of favorite foods and planning extensive menus.

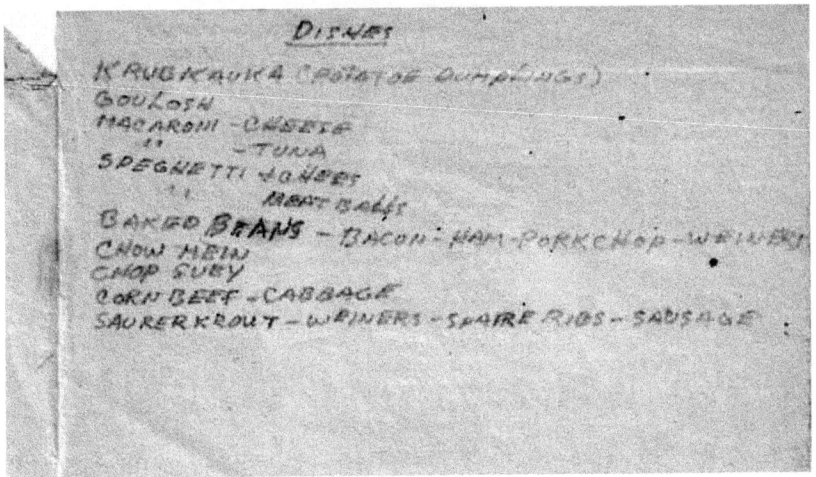

My precious list of "dream foods"

I carefully considered what I wanted for my first meal upon my return home. Others did the same and sometimes we shared our lists. It was interesting to discover the different foods that were favored by men who grew up in different parts of the United States. Chocolate was generally close to the top of the list everywhere.

WWII Prisoner of War - How I Survived

```
DESERTS                                          COOKIES
GRAHAM CRACKER PIE    ANGLE FOOD CAKES           BROWNIES
FRENCH APPLE  "         DEVILS    "    "         TOLL HOUSE
BLACK BOTTOM  "         MARBLE                   ICE BOX
PECAN PIE               POUND                    OATMEAL
FRIED "                 FRUIT                    HERMITT
BUTTERSCOTCH "          SPICE                    P. BUTTER
BOSTON CREAM  "         SHORT                    NUT
BANANA  "               COCOANUT LAYER           MACAROONS
COCOANUT  "             PINEAPPLE                COCOANUT
CARMAL  "               SPONGE                   VANILLA WAFER
CUSTARD  "              ICE BOX                  CHOCOLATE
APPLE, CHERRY ETC  "    BANANA                   ROCKS
MINCEMEAT  "            NUT                      FIG NEWTONS
FRUITCAKE  "            MARSHMELLOW              GINGER
PRUNE  "                UPSIDE DOWN
RASIN  "
CHOCOLATE  "
CHERRY  "               CHOC ECLAIRES
BLUEBERRY  "            BISMARCKS
RASPBERRY  "            CREAM PUFFS
LEMON CREAM  "          DONUTS (ALL KINDS)
BUTTERSCOTCH  "         NAPOLEAN TARTS
PINEAPPLE  "            TURNOVERS
PEACH  "                DEVIL DOGS
SOUR CREAM  "           GONDULAS
HUCKELBERRY  "          CHARLOTTE ROSSE
PUMPKIN  "              SWEDISH PASTRIES
APRICOT  "              DANISH  "
BLACK BOTTOM  "         CREPE SOUSETTE
TAPIOCA PUDDING

CUSTARD
FRUIT JELLO
PRUNE SOUFFLE
CARMEL
CORN STARCH
CHOCOLATE
JOHNNY BULL
PLUMB
RICE (RAISINS)
COCONUT CUSTARD
BUTTER BRICKEL
GRAHAM CRACKER
KREML
PORCUPINE PUDDING

DEVIL FOOD CAKE - PLAIN HERSHEY - COVER WITH CHOCOLATE PUDDING
FRIED BANANAS
  "   PINEAPPLE

BLACK BOTTOM PIE: - CHOCOLATE FILLING - COVER WITH
            VANILLA WAFERS - CREAM FILLING - ETC.
PORCUPINE PUDDING - SPONGE CAKE SATURATED WITH FRUIT
  JUICE (ORANGE) - STICK IN PECANS - TOP WITH
  WHIPPED CREAM + CHERRIES
```

Another day, different list of "dream foods" (written on another part of the same paper because paper was hard to come by)

Every day was flooded with new rumors and speculations about the war; many of which, we agreed, had their origin from the latrine. This much was clear, however, the Germans were gradually retreating on all fronts and the Allies were closing in from both the East and the West. Fighting was fierce in all sectors and it seemed clear to us that the Allies were winning. One day, in the coming few months, the war would be over and we would be liberated. Two burning questions were debated endlessly: when would this occur, and more importantly, who would liberate us? Would it be the Americans or the Russians? We passionately hoped that it would be the Americans.

Stalag Luft 3 was located in the northern part of Germany and it was very possible that the Russians would get to us first. We feared this possibility with an absolute passion. We had heard many reports about Russian brutality and we firmly believed that if the Russians captured us we would simply become their prisoners. They would load us on boxcars and ship us to a slave labor camp in Siberia. This seems like an extreme statement to make regarding Russia, our war-time partner however, I believed that scenario. The Russians were tough troops and brutal beyond the experience and the imagination of most Americans. We feared we would simply disappear and never be heard from again.

The Germans were in absolute charge of the camp with their guard towers and triple barbwire fences and the sentries, guards with dogs, and ferrets[5] all on constant patrol. However, in order to minimize problems with the Americans, a large part of the internal camp discipline was delegated to the American chain of command. Our senior American officer ordered that every man march ten laps around the compound every day in order that we

---

[5] Ferrets are guards who inspect the crawl space under the barracks looking for evidence of escape tunnels.

all stay in shape. Few other commands were given and we had considerable freedom to go and come within the camp itself.

As I had learned the hard way, the krieges of Stalag Luft 3 were a well informed group. We received the official German war news every day. We also received American and British news through an elaborate and ingenious secret radio set-up that had been developed by previous krieges over the preceding months and years. The secret radio had come into being through resourceful ploys that were quite incredible. There were skills present to build and do most anything with two thousand intelligent, educated and resourceful men in camp.

I learned of the creation of the secret radio and was fascinated by its story. At one end of my building block was a small room that housed a man who had been in the camp a long time and who spoke German fluently. This kriege, working with the backing and resources of the X Committee, carefully crafted a friendship with one of the German guards. The guard's family lived close by and he saw them regularly.

The kriege chatted with the guard in German whenever he could. Sometimes, on cold days, he invited the guard into his room to get warm and to have a cup of hot chocolate. During this time the guard would help the kriege with his German. Occasionally the kriege gave the guard a chocolate bar that had come from a Red Cross parcel to give to the guard's daughter and family. As this happened a number of times, trust was gained as other gifts and favors were given to the guard and small favors were returned. This process went on for some time.

Then one day, a couple of months later when everything had been properly prepared, the guard was called into the room with the kriege. Members from the X Committee were waiting there. The guard was then informed that every conversation and gift had been observed and logged. He was told that this record would be turned over to the German Gestapo unless the guard cooperated in getting a few small items such as thin copper wire.

The guard was now facing a crisis. If his record of collusion with the Americans was reported to his superiors he could be charged with treason and treason was punishable by death. The guard had no option except to cooperate with his American friends and come up with the radio wire. This process, of course, gradually expanded to include larger items that were needed such as radio parts. The more the guard cooperated the more he had to cooperate: war is never fair.

The reports from the three sources of news were not always the same. In general, I preferred the British news because it was the most accurate. The German news was usually late and distorted as they tried to cover up their defeats and expand their successes. The American news was eagerly received but was generally overly optimistic regarding Allied progress.

There were a few high moments in camp having to do with entertainment and sports that had been arranged. The most uplifting and notable event of all occurred on Christmas Eve of 1944. Great efforts had been made throughout the camp to bring the spirit of the Christ into our midst for the occasion. Special Christmas worship services were led by a chaplain and a Christmas play was presented. Later in Room 13, we celebrated by holding our own special Christmas party. We had prepared for the party for over a month. On Christmas Eve we basked in an aura of fellowship and goodwill as we gave gifts to one another. We had drawn names and each of us gave a gift to the person whose name we drew. I spent a good deal of time and effort in carving a wooden cigarette holder with an Air Corp insignia for the person whose name I had drawn. I received a pair of suspenders that had been carefully woven from hard to get brown paper and they actually worked pretty well.

We sang Christmas carols and even danced to the music of the single record and only phonograph that existed in camp. This Christmas party still stands out as one of the special moments in my life. This warm evening of fellowship was a stark contrast to the grey depression that marked our usual routine. For me the spirit of Christmas had entered the picture and changed things for one memorable night.

Chapter 13

The temperature dropped and winter settled in. It was to become the most severe winter on record in that part of the world. We continued to discuss food and to avidly watch the progress of the war through our news sources. We grew more nervous about our future as it appeared that the Allies were winning but the Germans still had a lot of fight left in them. It seemed to me that the tone of the fighting was becoming ever more desperate and vicious. I worried that a showdown was developing and that we might be caught in the middle of the fighting. My sense of nervous uncertainty increased.

Then in the middle of January 1945, the Russians began their long awaited assault from Warsaw toward Berlin. By studying our maps, it appeared that we were in the middle of their expected path. This worried me because I greatly feared being liberated by the Russians, being shipped to their slave-labor camps in Siberia and laboring in the Siberian mines for about six months, which was the typical life expectancy. This is an idea that I still hold today.

The bleak afternoon of January 28, 1945, had an ominous feel to it. With the rolling rumble of Russian artillery in the distant background, my friend Hal and I, along with hundreds of other men, dutifully hiked the mandatory ten lap conditioning march around the camp perimeter. I was trying out the shoes that I had just acquired from Hal. He had been able to get a different pair of shoes and he gave me his old ones. His broken down shoes, size 11 EEE, were several sizes too large for me. Nevertheless, they were a huge improvement over the small ill-fitting ones that had been issued to me. This seemingly insignificant event of getting a different pair of shoes is one of the two or three important incidents that enabled my survival of the war. I am fully confident that without this change in shoes I would not have endured what was yet to come.

The nervous and uneasy tone in the camp deepened as the day worried on. In the cold early evening, Hal and I hiked over to the theatre building to take part in the choir practice held there. We entered the room and observed that less than half of the members were present. The room was nearly empty. The men who had assembled were distracted and uneasy. Our edgy choir director paced up and down with uncertainty trying to make up his mind on whether or not to call off the rehearsal. After a few moments, he made his decision and abruptly declared that choir practice would be canceled for the night. With relief we headed out into the night and back to our room. This was not a night to sing.

The door to Room 13 opened to a scene of grim men silently and anxiously making final adjustments to their personal equipment. A quiet feeling of foreboding and apprehension was in the air. Outside the weather was ominous and the temperature was dropping. A huge blizzard was starting to move in to grip the area. In the distance, the Russian artillery flashed and rumbled.

I joined the worried activity of checking equipment and getting ready for a possible forced march on a cold night toward somewhere. Suddenly the door flung open. Someone stepped in and breathlessly shouted, "Room Fuehrer to Major Ott's room on the double"! A collective gasp could be heard. This was it: we would shortly know our destiny.

Hal, our Room Fuehrer, dropped his pack, rushed out the door and headed toward Major Ott's room to get information and orders. The waiting during the next few minutes was endless. Suddenly we heard the clatter of boots pounding down the hall. Our door banged open and Hal rushed in wide-eyed and breathless.

"We assemble at the front gate," he shouted. "We march in one hour. Carry what you can, destroy everything else. Don't leave anything for those damn goons. The Russians could be here in the morning."

The room burst into frantic but purposeful activity. The food hoard that had been so painfully collected over the months was opened. It was hastily heaped on the wooden table in the center of the room. Each man took a food packet that had been previously sorted and jammed it into his pack. Life had suddenly become clear. We would march now, tonight, under the Germans and ahead of the Russian advance. This was both good news and bad. It was good because we would not be turned over to the Russians. It was bad because we would march in cold weather without proper equipment and at the edge of what might become an infantry battle. It was also good that the Germans wanted to keep us to use as possible bargaining chips if a time for negotiations came up. However, it was also bad because the Nazis would not hesitate to sacrifice some, or all of us, if they thought it useful or necessary for their well-being.

Our packs were a marvelous collection of ingenuity and inadequacy. Each was unique and crafted by its owner out of what few materials could be found. I had made a backpack out of a cotton summer shirt. The shirt fit on my back so that the buttons opened to the rear, with the sleeves arranged to drape over my shoulders as straps. It was a sensible and workable pack; however, I had failed to test it properly to discover its flaws. In actual use it turned out to be uncomfortable, small and flimsy. I took care to balance my pack as best I could. It contained my blankets, my portion of food from our acquired pantry and very little else because I was wearing everything I owned. I was in about as good a shape as I could hope to be in this situation. I was wearing the two pairs of long woolen underwear which I had bartered for during the preceding weeks. I wore my two pairs of cotton socks and the very old, completely broken down hobnailed size 11 EEE boots that I had acquired from Hal. These boots were much too large for me, but they could accommodate the cardboard inserts I had made. The inserts slightly improved the fit of the shoes and also helped to make them a little warmer. My serious lack was that I had no gloves. I had made a pair of gloves out of a cotton towel, but in below zero weather, cotton gloves are very close to no cover at all.

It was about nine at night when we were ordered to assemble at the front gate. We stood in the cold; two thousand one hundred men, milling around in the dark trying to keep warm as the temperature steadily continued to drop. Like me, everyone wore every bit of clothing he could find. All had prepared, as best he could, for entrance into a cold and uncertain future. The mood was nervous and grim. A difficult, dangerous and crucial episode was about to begin. There were no pleasant or optimistic options in sight. We were about to march, without proper equipment or organization, just as a savage blizzard was starting to take hold. To make matters worse, we were marching in an area that was likely to become a major battleground between Russian and German forces and toward a destination that had not yet been determined. Only the military could arrange a scenario as perfect as that.

The temperature dropped steadily. We milled in place, stamped our feet, swung our arms and gradually became thoroughly chilled while we waited for the word to move out. The German soldiers assigned to guard us were not yet ready. They needed time for preparation so we waited, and waited, for our guards to get ready and for the other four camps to move out. By midnight, the temperature had dropped a good deal below zero. We were chilled and finally downright cold. Shivering, with hands and feet that were numb, we shuffled around in the dark in a state of impatience, growing confusion and deepening fatigue. Finally the word was shouted down the line, "line up single file and march through the front gate. Pick up a Red Cross parcel. Move out and hurry. Rouse! Rouse"!

Finally, thoroughly chilled and stiff with cold, our mob of men began to move. I edged my way toward the front gate and the deep darkness beyond. I caught the Red Cross parcel that a German soldier tossed toward me as I passed by the supply shack. I tucked this awkward thirteen pound package under my left arm and trotted to catch up with the column. The food parcel was about fourteen inches square and six inches high. It contained carefully chosen canned food, candy and cigarettes. I was grateful to receive this parcel but it was a very cumbersome thing to carry. In my struggle to catch up with the column I had no opportunity to repack my load. Before long, I was having a hard time trying to manage the awkward package and my pack while struggling to catch up with the column. In a short time, I was drenched with sweat, my arms were aching and I was getting tired.

Our pace finally slowed to a hard walk, and after a time, to a steady walk through dark and heavy woods. Gradually the members of Room 13 became strung out and separated. Although we tried to stay connected, it became increasingly difficult. The wind continued to pick up as the cold deepened. Trees swayed and cracked. The snow was dry and crunchy. A solid Minnesota type blizzard was setting in and showing its muscles. Snow and ice particles whipped and whirled about making it difficult to see. The column settled into a steady walk in the icy darkness.

Doggedly hiking along a dark road an hour or two later, I found myself walking next to Lieutenant Warren. He looked miserable. His cumbersome blanket role was draped over his right shoulder while he clutched his Red Cross parcel in one hand and his greatcoat in the other. He was having a hard time, and like me, was sweating profusely in spite of the cold. We grunted in recognition of each other and labored along side by side. Finally Warren lamented, "Damn it, Len, this overcoat is so heavy and I'm so tired I think I'm going to throw it away."

Shocked at the thought, I responded, "Well do as you want but that would be a big mistake. You guys from California don't know what a blizzard looks like. I'm from Minnesota and have been in a lot of blizzards and the last thing I would get rid of is my overcoat. You're going to need it before this thing is over. You would do well to keep it."

"Maybe," Warren responded, "but I'm so awful hot I can hardly stand it. I think I'm going to throw it away."

"A blizzard is coming up," I countered, "and I will throw away everything else before my overcoat goes."

He grunted with indecision as we continued to hike together in the wind and swirling snow until we lost track of each other.

The column continued on into the night. At one point, as we passed through heavy woods, we waded through a small stream in which the water was eight to ten inches deep. Our shoes filled with water and our legs became wet to the knees. Our feet squished with water and our pant legs were soaking when we finally achieved higher ground and began to walk on a road again. At one point, as we made our way over a hill, we passed by several squads of German infantrymen dressed in white winter camouflage busily setting up machine gun outposts. They appeared to be preparing their defenses to face the Russian assault they were expecting.

With great effort, dawn began to appear over a gray horizon. The temperature had now dropped to about fifteen degrees below zero and it was steadily getting colder. Clearly a major blizzard was preparing to unleash its full fury. I grew up in Minnesota during the cold winters of the 1930s and I also lived in Montana for fifteen years. In all that time, I had never seen a winter storm as severe as this.

My roommates dragged along, still more or less together, with Hal and me somewhere toward the rear. Our lack of rest and food and our all night exertions were taking their toll. All of us were having a difficult time. Passing by an opening in the forest, Hal and I saw a man floundering in the snow at the side of the road. We recognized him as Castleman, one of our roommates from West Camp. We stepped off of the road to help him. His problem was obvious. His oversized pack was far too heavy and cumbersome. He was carrying way too much and he was floundering and becoming incoherent. Against his mumbled protests, we dragged the pack off his back and sorted through it. We threw away heavy canned foods, as well as, other items that we thought were not essential. During this procedure, we berated him, cussed him and encouraged him. By the time we had finished rummaging through his pack, he had begun to recover somewhat. We put his pack back on him and the three of us returned to the road and the struggling column.

Walking became increasingly difficult and our vision obscured due to the increasing winds and the blowing snow. Gradually the column strung out and any semblance of leadership disappeared. Any sense of organization slowly evaporated as we gradually became a mob of individuals concerned only with personal survival. We were a muddled and confused mob of men with no idea as to where we were or where we were going. We found out later that our leaders were as confused as we were. Apparently, there was no real plan as to where to house what was left of 2100 exhausted men. We were simply blundering along seeking refuge from the storm and the Russians.

The storm continued and even increased in its fury. Visibility decreased as the temperature continued to drop. Men were staggering with exhaustion and some were lost in the blowing snow. A couple of times gunfire broke in the night as the guards fired at someone or something. Now even the guards grimly struggled to stay alive and to continue on while the storm screamed forth its furious rage. Most of the guards who started with us were middle-aged men. Some of them had dropped out while a few sat on wagons pulled by horses. The older guards were having a hard time and soon the top priority became the same for everyone: survival.

Hal, Larry, Norman and I, all from Minneapolis, Minnesota, paused to share some food as dusk approached. We made a futile attempt at building a fire but were unsuccessful in getting it lit. As men who grew up in winter country, our speculation was that the temperature had dropped to around forty degrees below zero. We did not know the actual temperature but it passed the scientific temperature test that I had devised as a fourteen year old boy with a morning paper route. The test is simple. You simply conjure up a wad of saliva and spit. If the saliva hits the ground as ice, it is at least twenty-five degrees below zero.

The march had become very difficult. We needed rest, food and shelter and it was simply not to be had. Men who were having trouble hanging on began to lighten their packs by throwing things away. Our situation had progressed beyond difficult and was approaching desperate. I found myself next to Warren again. He was wobbling along all wrapped up in his blankets. He had thrown away his woolen greatcoat. I thought to myself, the damn fool did it. I'll bet he wishes he had that greatcoat now. I gave his problem no more thought as I was in too much personal torment to think about him or anyone else. Everyone was having trouble. Some men had thrown away most of the contents of their packs to rid themselves of weight and a few were dragging along without packs. Desperate fatigue was affecting everyone.

Some of the German guards had been rotated but a few of the original guards were still with us and they were in almost as bad a shape as we were. I recall seeing a fifty year old German guard staggering along in the midst of the column and dragging his rifle along the ground by the strap. This would be an offence worthy of court martial in ordinary times but at this point he became one of us and was only trying to survive.

Around dusk we labored across an open space. There I saw an innocent little scene that will be forever etched in my memory. I looked toward the field on my left and there, barely visible in the midst of the driving snow, I could see a small two-wheeled cart with slanted sides: one that could have come out of the Middle Ages. There, harnessed to the cart, stood a totally exhausted horse. He was standing perfectly still with his legs spread apart and his head hanging down. He was obviously completely exhausted and ready to drop. I was told the horse actually did drop a short time later. The cart was filled with straw and household goods. Seated on the front of the cart and driving the horse was an old man. He was heavily bundled against the howling wind. In the back of the cart sat an old woman, probably his wife, who was also bundled against the cold. In her arms, she cradled an infant child to protect it from the unrelenting weather. For me, this little scene summarizes the reality of war: a dying horse, a defeated old man and a desperate woman clutching a helpless child.

Chapter 14

The unrelenting cold continued unabated as my part of the column entered the village of Freiwaldau at dusk. Word came down the line that this was to be a rest stop and that every man would get twenty minutes rest and food in the local jail. I stood outside in the driving wind for about an hour until it was my turn to enter.

The heat struck almost like a blow as I walked into the warm building. I became faint for a moment. Then, as I passed by a counter, I was given a cup of warm water. I sipped this delicious liquid slowly and lovingly. It was wonderful and exquisite beyond expression. Carefully nursing my warm water, I made my way over to the side of the room and laboriously sat down against the wall to savor the wonderful fluid to its last drop. My pack struck something on the floor, and as I tried to move, I realized that I had sat on something the size of a grapefruit. I gulped down my water and reached down to push the impediment away. Low and behold, I found a tightly wound ball of socks that some previous kriege had unwittingly left behind. This was absolutely a godsend for me.

Good socks were something I desperately needed. I was astonished and grateful to find just the thing I needed. With this new gift for survival, I was stimulated to stay in motion and to prepare for the outside cold that I would soon again be facing. I tried to ignore my fatigue as I forced myself to unbutton my overcoat, take off my shoes and rub my feet. I replaced my socks with the three dry pair from the bundle I had found. I also replaced the cardboard inserts in my shoes. I continued my efforts to ready myself by sorting the things in my backpack to make it as balanced and comfortable as possible. After I finished my preparations I leaned back to close my eyes and sleep. However, the infallible Murphy's Law prevented the possibility of sleep. Orders were shouted that we should leave the building to allow for other men.

It was a shock to walk out into the darkness and face the savage weather. The temperature had dropped even further and the wind had increased in the brief time that I had been in the warm jail. We were ordered to line up in columns of three and prepare to march. Wind blew ice particles parallel with the ground as we started out. I could hardly see the man three feet ahead of me as we began our agonizing march into the night. The hours ahead were as cold, anguished and painful as anything I had ever experienced. Moaning and cries for help were a constant accompaniment to the howl of the wind. Each man withdrew into himself and gave in to his own desperate pain and fatigue.

Somewhere in the middle of the night, I began to consider giving up: to simply sit down and let the rear guards shoot me when they found me. I stopped walking and simply stood. I did not sit down because I knew that if I sat down I would not have the strength to stand up again. So I stood dumbly and silently at the side of the road as the ragged column of spent men dragged past me.

Finally, with a fatalistic attitude of - to hell with it, if they shoot me I don't care - I decided to sit down. As I made that decision, I realized that I was now at the end of the column and that several German guards were bringing up the rear and were beating fallen men with their rifles. I stood dumbly and watched them as they came closer and tried to decide what to do. Finally, when I was the next man in line to be beaten, I turned toward the column and forced myself to hike again.

My feet hurt terribly. With each step I felt a stabbing pain that was nearly unbearable. I soon discovered that if I walked where the snow was deep, soft and unbroken my feet did not hurt nearly as much. I found that the pain became even less if I walked as rapidly as I could. So I increased my pace and walked on the side of the road where the snow was heavier. In the process, I seemed to gain strength.

I began to pass the staggering men in the column with a newfound energy. I recall passing by someone and heard him mutter something to the effect that I was an idiot for spending my strength walking at such a pace. This process continued until I again found myself somewhere in the middle of the column, staggering and moaning with the rest.

The wind and the cold began to moderate somewhat a few hours before dawn. The full savagery of the blizzard was beginning to pass. I hoped that rest and shelter would come along with the new day. I was fully spent. Gradually I became gripped by an inner certainty: I was dying and I would not survive to see the dawn. For the first time in my life, I knew with an absolute certainty that I would die. My death would occur within the next few hours.

I was dying. I did not want to die and I was afraid to die. Terror engulfed me. Then in a state of near panic I began to pray. I prayed with desperation. "God…. let me live." I prayed fervently and, to make my prayer more enticing to God, I bargained, "let me live and I will go into the ministry."

I continued to pray this prayer many times. Time passed and somehow I found strength to painfully put one foot ahead of the other and do it again and again. I continued this process endlessly through the tortured hours until dawn began to suggest its coming.

The word from the front of the column was murmured and mumbled down the line, "Three kilometers to Muskau. Food and rest in Muskau." I thought to myself, I can make three kilometers. I cannot make four kilometers but I can, somehow, make three. And so, along with the others in the moaning column, I labored on. Soon the word came back that it was five kilometers to Muskau. Five kilometers, I can make five, not one more, but I can make five. And so it went for the next hour or two until finally we bumbled our way over a ridge. Below us was nestled the little town of Muskau. I can make Muskau, I told myself again and again, but I will not leave Muskau. I'll find someplace there to stop and rest, but dead or alive, I will not leave Muskau.

Now the goal was in sight and I, along with all the others, limped and staggered along toward this final hope for survival: Muskau! I would somehow find a place to rest here, but I would not, could not, go one step beyond. The need now was to somehow hang on for just a little longer. A new beginning for life or for death would soon be decided.

Chapter 15

The snow covered town of Muskau looked bleak and empty in the early dawn as we dragged ourselves into the town square. Our formation was completely disorganized. The bewildered and totally exhausted men straggled in and wandered in various directions looking for shelter. I noticed a three-story apartment building off to my left and made my way toward it. The main entrance was set back into the building a bit and I thought I would try to get through that door and lie down in a hallway somewhere inside. Other men were thinking the same thing and several of us converged on the doorway at the same time. As I neared the front of the doorway, I passed by a man in the street and noticed he was struggling to lift his foot up the two-inch curb. He was fully exhausted and did not have the strength to lift his leg. I numbly watched him for a while and then labored over to him and helped him step up over the curb. He staggered toward the same door that I had picked out. A number of men had also gathered near the doorway seeking entrance. I studied the scene for a moment and decided there would not be room in this place for me.

I looked across the central square for an alternative and saw Hal grimly dragging himself into town amid a string of men. I decided to intercept him on the other side of the town square and I angled my path and joined him just as the column turned up a side road. He wearily nodded to me in recognition and then grunted that there was a factory up the road where we might find shelter. Silently we joined the straggling column and continued on for about a mile until finally a three story red brick factory building appeared. We made our way around to a door on the side where we joined a gathering of men who were filing in.

The wall of 110 degree heat that hit me as I entered the factory building was almost like a physical blow to my cold and exhausted body. My legs sagged and I almost melted in its embrace. In fact, a few men actually did faint on entering into the heat. I followed the orders to climb the open wooden stairs toward the second floor of the building. Laboring up the stairs I noted that there were open racks all around that were stacked with raw clay plates and bowls. It occurred to me that this building must be some kind of factory that made bricks or clay products. This explained why the building was so warm. We had stumbled into the perfect spot for our needs: a place that was warm and dry.

This is a typical warmish, dry shelter that we flocked to on the second march *

The top of the stairs opened onto a large flat area above the kiln and it was covered by a layer of sand. Hal and I looked around to claim a place where we could spread out our blankets for sleep. We found Norman Grant and Larry Borsheim, our roommates from Room 13 at West Camp. The two of them were laying out their blankets on the warm sand. There was a bit of room next to them so Hal and I claimed that space with our blankets. We put our packs to the side and dropped onto our blankets. Within moments I was in a deep dreamless sleep.

It was late in the afternoon when we were awakened by the guttural shouts of the guards and a general commotion on the floor below. Hot soup had been brought in and it became necessary for us to get organized so that the soup could be ladled out in an orderly way. We each got out our aluminum bowl and joined the food line. My body was stiff and sore and objected to every move but I was eager to get hot food.

The soup was thick, nutritious and superbly delicious, without a doubt the most special soup that I had ever eaten. The nourishment of the soup helped to perk me up a good deal, although I remained too stiff, sore and exhausted to move any more than was necessary. We heated water over the vent from the oven below us and savored each delicious and precious drop. Even with the five or six hours of sleep that I had, I was still exhausted and it was not long before I again found myself stretched out on the warm sand deep in sleep for the night.

The next morning began with the strident voices of new and fresh German guards asserting their authority and waking us up with shouts of, "Rouse, rouse." We were ordered to get up immediately and prepare to march in half an hour. Our next destination was the town of Spremberg about twenty six kilometers away.

I was still stiff and achey and half-exhausted, as was everyone else. Nevertheless, we packed our gear and filed out of the building, urged on by the commands of the fresh guards. We were each given a slice of bread and a piece of raw potato as we left the building. Evidence of the fierce winter storm of the past few days was present everywhere. The temperature, however, had risen a good deal and the sky was rapidly clearing. It appeared to me that today's weather was going to be clear and pleasant.

Our column finally formed to begin the twenty six kilometer march to Spremberg. I was still worn out and my muscles objected to being used; however, there was no choice except to march as ordered. We had gone only a short distance when trouble emerged. I was having trouble as my legs remained stiff and it was a major effort for me to walk. I had pulled muscles and I was far more fatigued than I had realized. My buddies were having similar problems and this was not a good way to begin a twenty-six kilometer march. The whole column wobbled with pain and fatigue as it dragged along. We had only traveled half a mile when men began to collapse and fall by the wayside.

Our guards stopped the column and, after a half-hour of indecision, orders came that we should about-face and return to the factory. The proposed march would be postponed until the next day. It was obvious that few, if any, men could have made it. We willingly turned and gratefully labored the short distance back to our warm factory to collapse in its heat. The four of us from Room 13, Hal, Norm, Larry, and me, scrambled to claim a warm place together on the sand above the kiln. This began our formation as a combine: a group that would stay together, share food and watch out for each other throughout the rest of the war.

We were young men and, with the resilience of youth, the next few hours of rest and food brought about our considerable recovery. Mid-afternoon found the four of us wandering around among the other men trying to find friends or news regarding our buddies from Room 13. I hoped to discover something regarding the larger picture in which we were involved. I discovered that apparently our part of the column had made a wrong turn and had marched several miles farther than the men in front.

One thing that became clear was that we had lost a good many men and in later weeks we heard numerous stories about lost friends and roommates. I recall speaking to one man who was the sole survivor among his roommates. No official records were kept; however, our guesstimate was that less than half of the twenty-one hundred men who started the march had made it to Muskau. After asking around, it appeared that we were the only room from our block back at Stalag Luft 3 that had completed the march intact.

By late afternoon I was feeling much stronger and decided we could all use some lightness to break up the prevailing solemn atmosphere. Hal and I found a couple of guys from among our friends at West Camp and we formed an impromptu barbershop quartet. We sang as many songs as we could remember and others joined in. We sang in loud and boisterous voices just to show the damn Germans that they hadn't beaten us yet! As night-time drew near, I was grateful that we could spend one more night in our warm, dry refuge.

Chapter 16

The next day dawned bright and pleasant with the temperature just below freezing. We were not fully recovered but we managed to labor the twenty-six kilometers to Spremberg without incident. We arrived before dark and marched into a military base where they housed us in a large room that looked to me to be some sort of special gymnasium. Every bit of floor space was used in the crowded room but we were warm, dry and we had received a bowl of good soup.

We started to bed down for the night when a nattily dressed German officer entered the room. He strode over to the wall in the center of the room about ten feet away from where I had placed my bedroll. He stood on a stool and, in cultured English, shouted, "Attention gentlemen. I am here to read the afternoon news."

The huge gymnasium, jammed with hundreds of scrawny prisoners milling around, quickly became silent and fully focused. We were eager to learn what was happening in the war so we listened intently. The officer introduced himself and, in a clear voice and radiating an attitude of authority, read the German news from a paper that he carried. He read about the military action that was going on in various key places.

At one point he digressed from his reading and launched into an impromptu and impassioned monologue regarding the war. With deep conviction, he pointed out that, even now, German soldiers were valiantly fighting off the cowardly and inhuman Russians and were preventing them from entering Berlin. He pointed out that, in doing this, the Germans were really defending mankind, including Americans. He stated that the Russian menace would seek to control all of Europe and then the United States. He told us that it was in our own interest to support the German cause and then, with great passion, he invited us to join the German cause and fight the Russians: our common enemy.

A murmur of excitement rippled across the room because we knew that the German news was always biased and late. With this in mind we reasoned that, in actual fact, the Russians might already be in Berlin. This supported our perception that the Nazis were on the defensive and it might be possible for the war to end within a few months or even weeks. As the officer left our large room, euphoria wafted across the room. We had eaten, we were warm and dry, and the war news was encouraging. It doesn't get much better than that. I stretched out in my space among the crowd, near my three friends, and went to sleep.

The next morning began with confusion. We were marched a few miles to a marshalling yard where a long string of forty and eight boxcars were waiting. With cursing, threatening and swinging of rifles, the German guards ordered us to board the boxcars. When one car was jammed full, a guard slammed the door shut and turned the bolt. The next car was similarly loaded, the door shut and bolted. Following a menacing motion by a German guard, I crawled into a boxcar with Hal, Norm, Larry and other men. We stood in the dark as more men crowded in. Finally the guard decided that the car was full and the door was slammed shut. Suddenly it became dark except for a tiny amount of light coming from two small openings on each side of the car. It was difficult to see much more than the dim outline of the men present. The atmosphere became grim and frightening. We did not know where we were going or how long we would be on this train. I was well aware of the fact that the Germans had jammed many thousands of people onto trains like this and many of those people had died. I was frightened and worried that this could be our fate as well.

The amenities of the train consisted of the two small openings that let in a bit of light - that was the extent of our luxuries. There was no water, food or toilet available and we were tightly confined shoulder to shoulder. Soon the train began to move and, as we jostled along in the gloom, someone shouted that we should get organized and start by counting off. The count began: one, two, three and on to sixty-one. Sixty-one men and their packs were crammed, shoulder to shoulder, on a small, dark, locked forty and eight boxcar.

There had to be some way of accommodating everyone but how? Various plans were shouted out by individuals in the car but none were really heard or even possible. We ended up seated on the floor with our arms and legs all tangled with one another. When one person moved, everyone around had to adjust in a process that spiraled out from the point of initial movement. The whole car became a squirming motion of misery. Tempers grew short. Cursing was inventive. Sleep was unattainable.

I recall one scene that occurred during the middle of the night. My legs were tangled with a couple of other men and my back was pressing on someone behind me. One of the men cursed me roundly for intruding on his space. I in turn cursed him back. He retorted, "Kiss my ass, Kovar."

"Put a mark someplace, you look all ass to me" I angrily shouted back.

Each hour of that endless night was absolute misery. The torment went on and on during the hours as we sat in our own urine and that of others. I feel at a loss for words to describe the utter agony of this experience but, if there is a pit in the abyss of hell that is the absolute epitome of misery, this experience would be a candidate for the bottom rung.

The train traveled on through the night. It started and stopped and occasionally switched tracks or sat on a siding for awhile. From time to time we would get reasonably settled but, before long, someone would move and the squirming and torture resumed. The discomfort and pain became worse as the hours dragged on and, by dawn, we were all near our wits end. Our condition was off the charts on a misery scale of one to ten. The rising sun brought a lovely clear day with it but none of us saw it, nor had we the eyes to appreciate it. Our burning question was: when would we get out of this intolerable boxcar?

I am grateful that my memory is fuzzy about the greater part of this train ride. Somewhere in my head however, I have a scene of the train stopping and our being allowed outside for a short time. Although I am foggy about many of the details, I do remember that at some point, we were shuffled around to different boxcars and our situation eased a good deal. We had fewer men on the boxcar and, while things were still pretty tight, they were more tolerable.

I recall another scene, on our second day on the boxcar, of being seated near the door that had been left partly open. I was carefully cutting four slices of bread, one for each of us in my combine. Larry watched me intently and then complained about my holding the bread with hands that were filthy dirty. I recall sarcastically asking him if his filthy hands were any more lily white and clean than mine. He looked at his hands for a moment, shrugged and said nothing. Dirt and grime was equally distributed among us all.

A clear day such as this brightened our situation somewhat but it also added an extra worry, the possibility of being strafed by our Allied fighters. Sure enough, in the late morning, a couple of Thunderbolts dived down on us from the front of the train. Their fifty caliber machine guns hammered and splattered at our train. Terrified, I steeled myself for a second pass but it did not come. My guess is that the fighter pilots in those P-47s had seen a Red Cross marking and decided that our train was not a good target for them. A record kept by one of our men stated that, in that particular pass, three krieges were killed and two were wounded in one of the cars up near the engine.

I am grateful that I do not remember most of that terrible trip, but it did finally end at dawn on the third day. The train came to a stop at a railroad siding. The sound of harsh German voices shouting commands could be heard along with the rumble and clang of doors being opened. Salvation was at hand. We had arrived somewhere and absolutely anywhere was an improvement. I, along with my miserable companions, waited impatiently for the door on our boxcar to be opened.

Nuremburg, Germany *

Nuremburg, Germany *

Nuremburg, Germany *

Nuremburg, Germany *

Nuremburg, Germany *

Nuremburg, Germany *

Chapter 17

Stepping out into the fresh air of a clear, brisk February day was a marvelous luxury that the power of words cannot convey. Even the angry shouts of the guards who lined the track were welcoming and invigorating. Following their orders we lined up in a column four men deep. I breathed deeply as I stretched and swung my arms. It felt good to move freely in the sweet, clean, delicious smell of fresh air.

Our combine, made up of Hal, Larry, Norm and myself, stood together near the front of the long string of motley looking men in front of the boxcars. We were being organized to enter a new place. I looked around to see what it looked like. Our new location was a large old and unkempt military housing area. It was surrounded by the usual triple-barbwire fences and guard towers. Over the front gates a sign was visible: Stalag 13D.

This was Nuremberg: a noble old city long known for its rich culture. Rumor had it that this camp had been converted to a holding area for military prisoners. Presumably it had just been emptied of Italian prisoners. Now it would house us. The camp was depressed, grim and dirty but it was also an organized and official place. I thought that it might provide a measure of security and order along with reasonable living conditions. In any case, Nuremberg was a welcome change to the uncertainty, danger and smothering misery I experienced during the preceding three days.

The camp had been laid out in an intelligent and orderly way. Barracks lined the sides of wide roads and a washhouse was located between every few barracks. It was surrounded by the usual barbwire fences, search lights and guard towers. The

Administration and supply buildings were located outside of the front gates. During preceding years, this camp had been used to house prisoners from many countries and places. After years of use it was now in a state of gross neglect, filthy dirty and it reeked with the aura of despair.

My combine was among the first groups of men who entered the camp. I marched through the gate with my buddies and down the road that ran through the center of camp. Guards counted off a number of men and pointed to a building as our line of krieges passed the barracks. Men designated in the count then peeled off and entered the building to claim a bunk. When my turn came to be counted, I scrambled into the building, along with the others, to claim my bunk. The building was dark, cold and grim. Having been smothered in a boxcar for a few days, I wanted to find a space to myself. In seeking privacy, I made a poor choice in picking out a bunk. Our combine of four had agreed to choose bunks somewhere in the middle of the barracks where it would likely be warmer. I also wanted to find a place that might afford some privacy if possible. With this in mind, I hastened to a lower bunk next to an outside wall. In my haste, I failed to notice that this single wall frame building had no insulation. I had chosen one of the coldest and most inconvenient places possible: a lower bunk on an outside wall. It was always colder near the outside and in February 1945 the nights were usually below freezing. I had made a poor choice and I had to live with it.

Being cold was a fact of life at Nuremberg and it led to innovative ideas to provide heat. I recall one morning, after we had been there a few weeks, hearing the crackling of wood being broken outside. I wondered what it was. When I went outside to fetch water from the single faucet in the washhouse, I was startled to see that krieges had ripped wooden boards off of its sides. Fear welled up within me as I wondered how the guards might respond to this wonton destruction of property. We could expect severe reprisals to follow this kind of sabotage. To my surprise however, there was no reaction from the guards. They simply seemed to ignore this destruction of property.

Washroom without walls Nuremburg, Germany *

This fact did not go unnoticed by us, and within a few days, a new industry was born - tearing apart the sides of the washhouses to use for firewood, bed-slats and other necessities. Soon most of the washhouses were stripped of all their wood siding. Only the roofs and plumbing remained intact. The fact that the Germans did not exact retribution for such destruction was puzzling. It was not like them to ignore discipline or the destruction of property.

Four conditions about life at Nuremberg during my two month stay stand out most in my mind: filth, bugs, hunger and bombs. The filth was the result of years of despair, degradation and neglect that had occurred before our arrival. The camp had already been in use for several years and, in time, the place had become dirty beyond description. We tried to clean up the camp, as ordered by our senior American officer, but it was difficult to do because cleaning supplies such as buckets, mops, soaps and the like did not exist. Water was also limited because there was only one water faucet for every one hundred and fifty men. Our facilities became even more strained as hundreds of men from various camps all over Germany were added regularly to our ranks.

Vermin were a basic feature of Nuremberg. They were always with us and there was no escape possible. They owned every nook and cranny. When we arrived, each bunk was already equipped with an old pallyasse[5]. When new and clean, a pallyasse can be quite comfortable. These, however, were old, lumpy, damp and heavy with mould and decay. In fact, they were a perfect home for vermin of every description.

I recall that first day when I tossed my pack on the floor and began to straighten out the pallyasse on my bunk. After I shook it into place, I felt something crawling on my arm. It was then that I discovered that it was virtually alive and moving with fleas, lice, bedbugs, ticks and other such forms of life. Looking at the pallyasse, I could almost see the writhing motion of their vibrant activity. To lie down on the bunk was to present oneself as a warm and tasty meal to these hungry hoards. There was never any escape from vermin in Nuremberg.

Another major feature of life at Nuremberg was hunger. It was a result of the fact that we subsisted almost entirely on limited German rations. The Red Cross food, that had previously subsidized our rations at Stalag Luft 3, had all but stopped. Late that first afternoon, after the barracks had become somewhat

---

[5] A 'pallyasse' is a large paper bag or gunnysack that is filled with shredded paper or straw. During WWII, a pallyasse served as a mattress.

organized, I lined up for my first meal. They served a soup that we named green death. We received this soup quite regularly. It was slightly better than another soup that we came to call grey death. We agreed that green death had probably been made from the boiled-up clippings from a local farm. Grey death however, had a different taste and seemed to include the run-off water from a barn.

Our lack of rations had become desperate when a shipment of food supplies from the International Red Cross arrived. I was excited and encouraged when I saw an old Chevrolet truck coming into camp loaded with Red Cross food supplies. I believe that it was these Red Cross supplies that enabled our survival because the German supplies were so limited. Since that time, I have always been partial to the Red Cross.

Unlike the limited food supply, bombs were bountiful at Nuremberg. Nuremberg was a major industrial city and, as such, it was the location of several important Allied targets. The city was bombed about a half-dozen times during my stay there. I experienced the bombings to be both frightening and awesome to witness. One midnight, a week or so after we had arrived, I was startled into wakefulness by the wail of air raid sirens. A few minutes later their mournful sound was interrupted by a huge explosion that rocked our barracks. I leaped out of bed, startled, confused and terrified, along with everyone else. I could hear the distant boom, crack and rattling of German 88 anti-aircraft shells exploding their violence into the air high above us. The British were making a high altitude night attack and the German anti-aircraft guns were trying to shoot them down. It was awesome and engrossing to see the beams of the German searchlights frantically scouring the sky as they attempted to pinpoint a British bomber. I could see that it was difficult for the searchlights to locate and make a fix on the incoming bombers. When a bomber was located however, it was virtually impossible for the bomber to escape. As the searchlights got a fix on a target, the German 88 gunners would focus on it and fire away relentlessly.

It was extremely dangerous to be in a building if it was struck by a bomb. Therefore, during the bombing raids, we ran out of the barracks or jumped out of the windows and stood in shallow slit trenches that had been dug between the buildings. On several occasions I cowered in a trench in the middle of a dark night with about a half dozen other men. We tensely held a long wooden bench over our heads to shield us from the falling shrapnel. On a couple of occasions, the bench was hit with pieces of spent shrapnel as they fell around us.

The British also made low altitude night bombing attacks that I found to be unnerving. These attacks came suddenly and without warning. In the middle of the night, a British high speed Mosquito bomber would scream in at tree top level. The low elevation enabled it to stay under German radar detection until the very last moment. Seconds later, the huge explosion of a one thousand pound bomb shook the ground, rattled buildings and shattered the night. At about the same moment, German air raid sirens would begin to wail. The snarl of the Mosquito bomber would then fade into the distance and be gone. The quiet of the night would again envelope the area, while I and my companions sat up for awhile and waited to see if another attack would follow.

The American daylight carpet bombings in the area were also impressive and a terror and wonder to behold. First, German alarm sirens would begin to wail. A few minutes later, dozens of our B-17s appeared high over head, flying in formation at about twenty five thousand feet. Their con-trails trailing behind them marked their progress as they resolutely moved toward their drop point. When they arrived at their initial point of the drop, called an I.P., the bombardiers flipped the proper switch and their five hundred pound bombs would begin to fall. Three or four minutes later the ground around the target area was slammed and pounded by a string of explosions. Shortly after dropping their bombs, the planes turned to return to their home base and I watched after the con-trails marking their path.

During the raids, German gunners doggedly defended their territory by peppering the sky with exploding 88 shells. I could see black oily splotches angrily reaching to grasp toward the incoming planes. Usually the first anti-aircraft explosions hit low and to one side. Then, as the German gunners corrected their aim, the splotches of violence got closer to the incoming planes. Occasionally a bomber would be hit by a shell and stagger out of formation, sometimes on fire.

During all of this, I stood with the other men holding a long wooden bench over our heads. Sometimes we heard bombs striking in the distance and occasionally we felt the shake of the explosions. I recall one occasion when a bomb exploded near enough to our camp to buckle the barracks that was located about one hundred yards from where we cowered in our trench. On those occasions, I was grateful for that bench.

Thus, the days and nights passed for me at Nuremberg. There were moments when life was not all about dirt, hunger, bugs and bombs. One morning, I witnessed a minor little scene that impressed me greatly regarding the mindset and values that develop during war. Germany had fought a long war in Africa, the Mediterranean and most of Russia and, in the process; they had lost a good many men. They were now looking for new replacements. In order to fill this need, they were trying to recruit older men and even boys.

I had stepped outside our barracks just in time to see a handsome, well-dressed German officer standing outside the triple barbwire fence that circled our camp. He was talking to a group of about a dozen young fourteen and fifteen year old German boys. It was my understanding that this officer was pointing out to the boys just how grubby, ineffective and impotent we American soldiers appeared to be. We were, in fact, a dirty scrawny bunch of men.

Apparently he was trying to encourage and embolden the boys before they went to the front and joined the German troops who were fighting the American advance. What this Nazi enthusiast failed to point out to these young boys was that when they got to the battle front they would not be facing a scrawny bunch of unwashed, undernourished, ill-equipped, rag-tag men like us. Rather, they would be up against the fully equipped, well trained, battle hardened American First Division: The Bloody Red One. These young boys who were just entering puberty would have about as much chance of surviving such combat and returning intact as would a snowball in hell. But such is the glory of war.

Nuremberg POWs *

Chapter 18

An exhilarating rumor spread through camp in late March 1945: Patton was closing in. This idea was eagerly grasped, repeated and embellished. The air became electric with enthusiasm and the hope that we might be freed in just a few weeks. Another rumor also ran through camp and, for me, it was a much more sober and frightening one. Namely, that the Nazis planned to march us ahead of the oncoming Allied forces and hold us in a special redoubt near Switzerland. Our lives would then be used as bargaining chips by the Nazi leadership as a part of their surrender negotiations. Arrangements like this had happened in past wars and there was no reason that this kind of arrangement could not happen again with us. However, if such a plan was to work, we prisoners had to be alive and under German control. Therefore, for the moment, the Nazis wanted to keep us alive and intact.

The troubling question was: would the Nazis try to use us as hostages and offer our lives in exchange for their protection? I was encouraged by another possible option. The alternative possibility was that the Allies, under Patton, would catch up to us and liberate us. However this option was also worrisome, in that we could end up in the middle of a showdown. We could easily find ourselves in the middle of a battle, with the Allies on one side and the Nazis on the other. So rumors abounded and we oscillated between fear and euphoria. These thoughts were widespread among us when we were ordered to prepare to march toward an unknown destination. It was with conflicted feelings of foreboding and hope that I gathered up my pack and prepared to march.

April Fool's Day of 1945 found us marching out of our filthy, squalid, vermin-ridden camp and heading south toward Munich. There had been much indecision in the days leading up to our departure. We krieges were disorganized and uncertain but the Germans were as well. In the midst of the confusion of that morning, the German guards seemed to disappear for a time. They were simply not at their posts. During this interval, I sneaked into the headquarters building. In the few moments I was there, I liberated or stole several official papers, which I still have. I then rejoined my buddies.

It was mid-morning when our long column of scrawny rag-tag men marched out of the dirt, danger and despair of that terrible place. My new sense of enthusiasm and freedom was shared by all of the krieges. It was invigorating to get away from dirt and grime and breathe the fresh air of a new beginning. I was excited and uplifted: Patton was coming on strong and the war was entering its final stages. Of course the Nazis were tough and capable and had a lot of fight still left in them, but the tide had turned and I felt optimistic and hopeful. I knew that there was still plenty of opportunity to get killed but the war was winding down and Allied victory was in sight.

Hours later, while marching around a curve on a small hill that overlooked the city of Nuremberg, I looked back to see a sight that filled me with wonder and amazement. It was a picture of devastation and destruction of monumental proportions. The great and noble city of Nuremberg, once the pride of German culture, was now a huge mound of twisted rubble. The streets were cluttered with the debris of shattered buildings. The bleak grey of broken concrete was everywhere and it laid its dismal pall over everything. The skeleton of wrecked buildings jutted up out of the rubble like broken teeth. The cultural nobility of several generations had been destroyed and, incredibly, I had been present on the ground near the target area during a significant part of that destruction.

A new spirit of freedom enlivened me as our column hiked south along a gravel road toward the little town of Polling, about fifteen miles away. It was a lovely day and we were surrounded by pastures and wooded areas on both sides. In saner times, this would have been an idyllic and picturesque sight. However, there was a war going on and what we saw was not the beauty of the setting, but rather, our P-47s strafing and bombing targets in the valley ahead. Smoke billowed up from the valley and marked the area for miles around in all directions. Our long column of rag-tag men slowly snaked its way toward the smoke beyond the plateau.

We had neared the edge of the plateau where it dropped down toward the valley below when suddenly a P47 fighter planes came rocketing up from the valley. The pilot had been strafing something in the valley below and, as his fighter plane snarled up over the hill in front of us, he looked down to see a long line of marching men. He no doubt thought that we were a column of enemy infantry. With that idea, he dumped the nose of his powerful fighter toward us and let loose with his 50 caliber machine guns. Immediately everyone jumped for cover as shells spitted down the road toward us.

I was terrified by this, more so perhaps, than by any other event during the war. This lethal assault seemed to be aimed directly at me personally. But, in an instant, the fighter had come and gone. Along with other men, I continued to cower in terror in the ditch by the side of the road. I waited for the fighter to swing around and again make another pass at us, but praise God, it did not happen. Possibly, the pilot recognized that we were not German infantry and he simply continued on to search for a more suitable target. In any case, he did not return. A rumor later came down the line to the effect that several men up front were killed in that single pass. Somewhat shaken, we slowly regrouped and our endless line of rag-tag men continued to march on toward Polling.

Second March *

Second March Casualties (face blurred) *

Chapter 19

We arrived in the town of Polling in the late afternoon and my combine and I claimed a spot in a barn for ourselves and our gear. We lined up to receive our food ration: a cup of watery soup, a thin slice of bread and a few drops of honey. The barn was part of a large and well kept farm. We were comfortable and dry and our Allied forces were heading our way. With this situation, my spirits rose along with a new sense of optimism - and for good reason.

I was now involved in a new start and no longer closely confined. I was happily free from lice, bedbugs and the constant crawling of vermin. The sun was shining and I was out in the open and feeling freer. I knew, of course, that we were still prisoners and that a war was still going on; nevertheless, things were looking good.

The dawn of April 5, 1945, brought with it another glorious spring day. As we marched south, I witnessed the awesome and terrifying sight of the final carpet bombing of Nuremberg taking place many miles behind us. The contrails of hundreds of American B-17s from the Eighth Air Force crisscrossed the sky on their way to bomb what was left of the rubble.

These bombing raids were carefully planned. The lead plane dropped smoke bombs marking the target for the bombardiers who followed. Then unopposed, the first wave of bombers bore in from the south to drop their bombs in a long string across the broken city. The next wave of bombers added another string of bombs in a line coming across the city from the southwest. The third wave dropped bombs from still a different direction. The bombs crossed the city like spokes in a massive wheel of destruction during the next two or three hours. Even though this carpet bombing was several miles away from us, I felt the shock and concussion of the bombs and I could almost smell the cordite. The Allied fighter escort that accompanied the bombers high above added their contrails to the unchallenged sky. The fighter boys were having fun playfully displaying their acrobatic skills. I watched this massive bombing as it pounded Nuremberg and wondered - if there was little left of this great city at our departure two days earlier, what could it be like now?

Mid-morning the following day found us slogging along in the midst of a steady rain. The road was slippery and thick with mud. My greatcoat was soaked and heavy. I was miserable and wet to the skin. Conversation was minimal and I felt lonely and depressed. A disheartening thought kept circling through my mind: no one knows or cares how we are or where we are or that we are. We are forgotten and unappreciated. This was certainly not true but it was what I felt. In my better moments, I knew that our families anguished over us and prayed for us steadily and that our entire nation had geared itself to bringing us home. However, what I knew to be true and what I experienced in myself and my feelings were quite different. I can remember that deep sense of defeat and depression even now. Wet, hungry, dejected and heavy with fatigue I slogged on with my comrades.

Several hours later, we trudged through a tiny village in the midst of a drizzling rain. We passed a picturesque little cottage that was bordered with a white picket fence as we made a sharp ninety degree turn into the village. A slightly built kriege was seated on a straight back chair in the middle of the lawn. He was staring vacantly into space as he absentmindedly gnawed on a hardboiled egg. Somehow he looked familiar to me. I studied him for a moment and then recognized him. It was Gould, our navigator from our original crew on Con Job. My depression vanished as I jumped over the fence and approached him.

"Gould" I exclaimed, "it's good to see you. How are you doing? You lucky jerk, where the hell did you get that egg? You were cozying up to some innocent fraulein, I'll bet," and on I went with my bantering comments, but he did not hear me. He only stared vacantly with unseeing eyes into space.

I chattered on for a few more minutes telling him how lucky he was that now he might possibly be picked up by the Red Cross and they might somehow ship him home to his wife.

"You lucky bastard," I said trying to cheer him up, "the war is over for you. You're going home," I said earnestly; although I knew what I said was untrue. He hadn't heard me anyway.

The column was moving on and I did not want to be separated from my combine so I bid him goodbye, jumped over the fence and hastened to catch up with my buddies. I had lied to Gould and I knew it. It was obvious that he had become a casualty of war. He was mentally and emotionally shaken and, in a real sense, he was no longer fully with us. Further, the odds of his surviving the war were not good. As it turned out however, he did survive the war and, although he was intact physically, he was emotionally shattered. He was a damn good man who became one of the many uncounted casualties of war. Every man has his breaking point and Gould had found his. No normal person who endured what he had experienced could remain unaffected.

The picturesque, almost medieval, town of Berching rose before us in the late afternoon of April 6. This was to be our billet for the night. It already housed hundreds of men who had arrived before us. Every man and his combine scrounged about to find an empty building, a barn or some dry place to bed down. Our part of the column ended up entering a Catholic church located near the town square. As we crowded into the church, my combine and I managed to find a section of a pew and the space under it still open and claimed it as our own.

More men kept jamming into the church however, and soon the floors, pews and altar were covered with men and their packs. It became crowded and tempers grew short. Hal called our little combine together to discuss a possible alternate plan for our housing that night. Together we decided that Grant, our golden glove boxer, would hold our place on the floor and the pew above it. Borsheim would try to conjure up some sort of meal from our small store of goods. Hal would go out and try to hustle some food, while I looked around outside for somewhere else for us to sleep.

As I made my way through the mass of men to the outside, I again became aware that the German guards were becoming less belligerent and more cooperative. We prisoners knew that Patton and his troops were closing in and the guards were also aware of this. In fact, Patton was coming on so steadily that we had begun to think it was possible for Allied forces to overtake us soon. This thought was also shared by some of the German guards and this encouraged a change in their attitude. As German soldiers they were, of course, duty bound to guard us but it was also becoming prudent for them to turn their heads and ignore us whenever they reasonably could. Our guards realized this was not a time to acquire enemies.

A lone guard outside the church watched me as I stood a short distance from him and surveyed the street. I was wondering what to do and where to go in order to find better sleeping quarters. The guard looked me over and then slowly turned and ambled toward the street on my right. With that, I quickly trotted left to the opposite side of the building and headed toward the center of town a few blocks away. The town square was surrounded by several three and four story stone buildings clustered around an old medieval well. I studied the buildings and picked out a two-story building to investigate. I went to the front and tried the door but found it locked. In going around the side of the building however, I found a partially opened window which rose easily as I pushed it up. I pushed open the window and crawled through.

Once inside, I found myself in a stairway that led to the second floor. I paused to listen. The building was silent and I wondered if I was lucky enough to find the building empty. Then I quietly climbed the stairs to the second floor. The top of the stairs opened to several office rooms. The door closest to me looked promising so I jiggled the doorknob and it opened easily.

A large empty room lay before me and it was flanked on all sides by several dozen large Nazi flags. The flags stood tall and impressive in their holders. This must be some sort of Nazi headquarters, I thought to myself. Then a second thought came; this could be it, our snug home for the night.

I retraced my steps back to the church, slipped past the guard who was trying to be invisible and entered the crowded church. It was now jammed with tired and hungry Krieges falling all over each other. I located my buddies, and a short time later, we spread our blankets out on the floor of the dry and spacious second floor Nazi headquarters. Things were looking up. We were living in grand style and the war was winding down.

As we left in the morning, we each took a Nazi flag for a souvenir. My three by five foot flag folded nicely to fit inside my shirt by day and it served as a blanket by night. I brought it home and have it to this day.

German discipline was beginning to dissolve as we resumed our march south. This new situation gave us a measure of freedom that we had not previously known and the column began to string out in a more leisurely fashion. Escape now became an easier possibility if we chose to consider it.

Late that afternoon Hal and I stood alone in a valley at the juncture of two roads. One end of the column was a block or so ahead of us and the other about a block behind. As we stood there, we discussed whether or not we should try to escape. Our plan was simple: just hike up to the top of a nearby hill and hide in the brush for a couple of days until the German forces retreated.

This plan was almost sure to work but we quickly decided against it. The big problem with our plan was that, even if we escaped the Germans, at some point we would have to go through American lines. We decided that the prospect of two men trying to go through American lines was much too dangerous to chance. At this time in the war, brutality was running high on all sides. We knew that it was dangerous to go through either set of lines, German or American, except in large groups - and even that was dangerous. Small groups of two or three men could easily be mistaken for an enemy patrol. In such a case, we could be shot before our identification became clear. Our first priority now was to stay alive. Since there was safety in numbers, prudence dictated that we stay with the big column.

A short time later we found ourselves hiking past a farm that bordered the road. I noticed the quaint, nicely kept farmhouse and saw that a window was open on the side of the house. Perched on the ledge of the window was a freshly baked pie that had just been placed there for cooling. The steam was still rising from it.

Our combine held a quick consultation and then sent Hal to knock on the front door of the home. As he performed a diversion, I sneaked around the side of the house and stole that gorgeous pie. A short time later, and half a mile down the road, we found a wooded area where the four of us shared the most delicious pie that has ever been made in the entire history of pie making.

The next day we continued hiking on the road leading toward Gammelsdorf. At mid-day the column stopped for a break and our combine inventoried our food store. We found a small can of liver spread with a hole punched in the top by the German guards. They did this routinely with canned goods so that the contents would spoil and could not be stored for escape activities. This particular can was quite old when we shared it. Mold had formed around the top. We scraped off the mould and then ate the rest of the contents.

In the late afternoon, as we approached the town of Gammelsdorf, my stomach began to roil uncomfortably and I began to feel sick. That night we bedded down in a barn that was part of a large farm but I did not sleep well. I grew increasingly nauseous, and by sunrise, I was vomiting and feeling extremely ill. I cared little that this was to be a down day and that the column was not going to march. I was too sick to care about anything. My buddies took me out of the barn and seated me in the sun on a wooden walkway that joined the farmhouse and barn. I leaned against the fence in the warmth of the sun but I could not eat, drink or even think. I felt terrible.

While lying on the walkway, I dimly noted that there was a flurry of activity coming up from the road. The event at hand was the presence of a very high ranking German officer, possibly a general. He was flanked by two or three aides and they were slowly walking toward us and assessing our situation. I was sick beyond caring and beyond expression but my misery focused on this monster that now represented to me the source of my pain and illness. As the senior officer walked by me, just a yard or so away, I kicked out at his leg as hard as I could. He saw my kick coming and deftly side stepped it. An aide quickly reached for his Lugar pistol but the officer raised his hand to stop him. After a brief pause, they continued their tour of inspection as though nothing had happened. Those around me were alarmed at what I had done. An attack such as the one that I had just made was more than ample cause to be executed on the spot. Lucky for me, this officer was apparently a mature man who recognized that I was sick and impulsive. He simply ignored the event and continued on his way. I am very fortunate that my demise did not occur at that moment.

Later that afternoon, I started vomiting uncontrollably and became too sick to function. My buddies tried to help me but there was little that they could do. They had heard that a Red Cross hospital truck was scheduled to come by and pick up wounded men at a designated farmhouse back down the road about a mile. Hal went to check for alternatives while Norm and Larry picked me and my pack up and half walked and half dragged me to the proposed Red Cross gathering point.

The designated farmhouse had been vacated and was now very quiet. Presumably, I was going to be the first of the sick and wounded to be picked up at this place. Norm and Larry dragged me into the barn, arranged a comfortable straw bed and placed water nearby. Having done all they knew to do, they left me there for the Red Cross truck. They then returned to the column. I was left alone and desperately ill to wait for the Red Cross truck.

Night fell. I lapsed into semi-consciousness and became deathly sick. Every cell in my body was toxic. I drifted in and out of consciousness during the ensuing hours but was too weak to help myself or to care. Death would have been a welcome release. Never have I experienced such misery and sickness. I lingered in semi-consciousness all that night, all the next day and the following night: alone, without water, often delirious and too weak to function.

When I awoke on the morning of the third day I felt extremely fragile but I had crossed some sort of threshold. I was weak and rummy but I knew that I was recovering. I realized that I needed to eat in order to gain my strength but I could not yet stomach food. I tried to find some clean water but, finding none, I settled for the little runoff stream that ran next to the barn. I spent the next few hours lapsing in and out of sleep. At around noon, I felt stronger and decided to gather my gear and make my way down the road to find the column and my combine. I was going to survive.

It was mid-afternoon by the time I slowly made my way back to the spacious and clean farmhouse where I had last seen my buddies. It was a different place now. The area was swarming with a whole new group of krieges in various stages of dishevelment. Across the road from the farmhouse was a large open pasture strewn with hundreds of men who were in every possible condition: wounded, sick, dying and dead. Apparently this had become the actual ambulance pick-up point rather than the place where Norm and Larry had placed me.

I slowly stepped around the wounded, sick and resting men and gradually made my way out into the field. There I found an open spot in the sun where I dropped my pack and collapsed on the damp ground. I lay there, rummy and half sleeping. I had slept for an hour or so when I felt a gentle kick in my ribs. I opened my eyes and looked up. Standing above me was the familiar face of Castleman, the same man that Hal and I had befriended while on the big march. He looked down at me and, in a soft voice asked, "What are you doing here, Kovar"?

"Hi Cass," I mumbled, "I've been sick - had ptomaine poisoning - feel better now though."

He studied me for a moment and then changed his tone to mock anger. "Kovar, you asshole," he snarled in a loud voice, "you look like green shit. God damn it, how could you let yourself fall apart like this? You're filthy! Can't you even shave? Shit, any fathead could do better than you"!

And so began a steady monologue that continued for the next hour while he got me up and moving. He brought fresh water to me and ransacked through my pack to feed me. He half carried and half dragged me across the road to the milk shed. There he washed my face, shaved me and got me going. All the while he carried on a steady stream of profanity and mock insults.

After he got me cleaned up and restarted, he led me across the road again to the pasture and found a place for me to lie down. He had me stretch out and I promptly dropped off to sleep. A few hours later, at dusk, I woke up. I was still quite weak, but I was feeling much better. I had turned a corner and could begin to function again. Castleman was gone.

I looked around and surveyed the scene of which I was a part. It was a picture of monumental pain and confusion. All about me, seated, lying down and standing, were hundreds of men in various forms of anguish and dishevelment. I sat there wondering what to do when a Red Cross ambulance truck arrived in the area and stopped for the wounded and sick. The truck was an ordinary open 4x4 truck and it was already jammed to overflowing with men. Some men were even hanging onto the sides of the truck. I started to walk toward the truck thinking I would somehow find room to climb on board and join them. As I got closer and saw the agony and desperation there, I suddenly thought, to hell with that…that is the end of the line and I want no part of it. I can make it on my own. Hoping that I could somehow find my combine, I turned to walk south in the direction of the main column.

I am not sure how I managed the next two days because I do not remember anything much except for a couple of vague scenes that mean little to me. I must have found something to eat and something to drink in order to have continued on as I did. I do remember, however, the morning of the third day when I caught up with the column and located the area where my combine buddies might be. It was early morning and the men were getting ready to line up and move out when I walked up to join them. They were taking a head count. A few men applauded when they saw me hiking up. They were surprised to see me alive and functioning. Those who knew me thought that I had died but here I was, alive and ready to join them again.

Chapter 20

The march to Stalag 7A, near Munich, was characterized by a big change in our tone and attitude. The war was winding down and, as it did, we all became more cautious but we were also optimistic. There was a sense of expectancy in our mood. We were feeling confident that the war would soon be over as evidenced by the fact that the Allies were now in complete charge of the air war. Allied fighters playfully waltzed through the sky unopposed. Our guards were becoming more considerate and cooperative, as were German civilians. It appeared that our release as prisoners would come in a matter of weeks or, at most, a couple of months. With the end in sight, it was too late to take unnecessary chances and risk getting killed.

Second Death march ends at Moosberg, Germany *

WWII Prisoner of War - How I Survived

Stalag 7A had become a catchall for prisoners from many nations. In this camp, one could find soldiers and mercenaries from all over the world. There were American and British soldiers as well as Scots in plaids and kilts. India's Gurkhas warriors walked proudly about with their turbans and flowing robes as did thousands of other men from everywhere. This camp had also become a distribution point for the Red Cross. This meant that, in addition to receiving the substandard German ration, we also received some aid from the International Red Cross.

Many things were in short supply during the time that I was at Stalag 7A. Food was very limited by any standard and we were overcrowded. I stayed in a section of the camp for American men and slept in an old tent that was torn and tattered. Since there were no beds or cots available I slept directly on the ground. There were no supplies available for clothing, medical needs or anything else but it was now spring and the weather was mild. All things considered, it was a steady setting.

Prisoners from many nations arrive at Stalag 7A *

Wild rumors raced back and forth across the camp daily and sometimes hourly. We knew that Allied troops were on the way and rumors ran wild stating that we could be back home in a matter of weeks. Our mantra was: don't do anything stupid, it is too late in the war to be getting killed. With that, I became more reluctant to take chances and I lived on whatever news I could get.

Moosberg, Germany *

Moosberg, Germany *

Moosberg, Germany *

Moosberg, Germany *

April 13, 1945 dawned bright, crisp and clear. It was a day that was alive with a sense of anticipation. This was a day that would be one of the most dramatic and empowering in my life. It was about ten in the morning and I was returning from visiting Norm in a tent that had been set up as a hospital. During the last march, Norm had acquired a blister on his foot and it had become infected. There was no way to take care of it properly and finally blood poison had set in. The proper medical procedure at the time would have been to apply hot packs around the clock and, hopefully, draw the infection to a head. If this was successful, the infection might then be drained away. However, in this primitive setting there were no medical supplies with which to perform the necessary procedure, so Norm sat on his grubby bedroll and worried as he hoped for the best and the infection in his foot became worse.

After visiting with Norm for a time, I left the hospital tent to return to my own. As I made my way back across the camp, I felt a tone of excitement in the atmosphere. A strange tension in the air had been increasing over the last few days and I could feel it now with increasing force. I had been expecting some sort of showdown to occur when the Allied forces finally arrived to liberate us. Now that time had come.

As I neared my tent, I became aware of an uncanny quiet that had enveloped the area. Something was different. It was as though the earth somehow stood still to acknowledge this special moment. Suddenly I realized that there were no German guards around. I crossed the old parade grounds going toward my tent and noticed that there were very few men to be seen anywhere. The camp, along with everything else, had become quiet. Even the air seemed to stop moving. There were no birds about or even bugs.

I stood still and listened to a strained, eerie silence: a silence that, in some sense, had substance and which I have never heard before or since. Suddenly the silence was broken and I was startled by the report of rifle fire coming from beyond the outer fence. Immediately, from within the camp came an answering shot.

They're here! The infantry is here and the German rear guard is firing back, I thought to myself. With that, I dove into a depression that remained from one of the slit trenches that had once crossed the field. The sound from small arms fire cracked through the air and I burrowed down even more deeply into my protective foxhole. Rifle fire erupted from two directions for a few moments. Gradually the gunfire became less and less frequent. Then everything became completely still. The universe itself seemed to pause. I held my breath and waited. Silence gripped all things.

I heard the distant rumble of a heavy engine, probably a Sherman tank getting into position. Patton is here! This is it, I exclaimed to myself as I hunkered down even more deeply in my shallow trench. The uncanny stillness deepened and nothing moved.

Then a strange and mysterious sound began to rise. It was a sort of murmuring whisper that powerfully rose to become an electric silence. I listened intently for a few moments and then, with caution, I carefully lifted my head above the edge of my foxhole to look around. Other men were warily looking out of their foxholes, seeking to assess the situation. Here and there a man was standing up and looking expectantly across the field toward the main administration building. I thought if it was safe for them to stand, it would probably be safe for me. I cautiously crawled out of the foxhole.

Standing up, I looked in the direction that absorbed the other men. An incredible sight came into view across the field and beyond the barbwire. The Nazi flag on top of the administration building was being lowered. I stood transfixed and silent in the grip of this scene as that flag slowly disappeared below the trees. Then, a moment later, another flag began to slowly rise on that standard. Gradually it lifted above the tree line to wave gently in the light breeze.

What I saw was the flag of my country, the Stars and Stripes. It was the American flag and it rose steadily to the top of the standard, strong and secure. But it was even more than that...it was my flag. Hungry, filthy dirty, scrawny as a rail and with tears running down my cheeks, I stood at attention and saluted - my flag. Never have I stood so tall, so strong, so straight, so proud....so humble.

Through the grace of God, I had survived!

Epilogue

The next couple of hours were marked by great confusion. Our camp was liberated by General George S. Patton's army. In the midst of it all, I left the camp on a supply truck. I rode on the truck for a couple of miles to a temporary air field which had been set up for forward observation. There I was able to bum a ride on a small Piper Cub observation plane that was heading for La Havre, France where American ex-POWs were being processed. I reached La Havre safely and was dutifully processed.

Finally, I boarded a ship headed for home. On the third day, while mid-Atlantic, the Nazi government capitulated and the war in Europe was declared to be over.

Officially, I was on leave at home in Minneapolis, Minnesota for two weeks before going to Florida to receive my next assignment. During my leave, I talked incessantly about my experiences as a prisoner of war. My mother, unknown to me, took detailed notes of everything I told her. When I left for my next assignment, my mother sat down, opened her notes, and wrote her version of my POW story. She recorded nearly all of it and gave it the title, "I Was There." She was amazingly accurate and complete in her record. I used her writings, along with my own personal records, in the preparation of this book.

We heard that we were going to be assigned to a new combat crew and take part in the final all-out assault on Japan with an expected casualty rate of about a million men. Before I was deployed, however, America dropped two atomic bombs on Japan and the war ended abruptly. I was discharged a few weeks later.

That fall I entered Macalester College in St. Paul, Minnesota as a freshman. It was there that I met Lorraine Bakke who would become my wife.

I graduated from Macalester College in St. Paul, Minnesota and married the girl of my dreams. I hoped to attain profitable employment and become wealthy. In fact, I spent the next year working for the University of Minnesota but my promise to God made during that brutal winter march never left me.

I decided to prove the ministry was not for me so I committed to one year at Andover Newton Theological School. Three years later I graduated and was ordained as a Congregational minister. I served in the Christian church for over fifty years.

The men referred to in this story survived the war: Mitch Cohen, Richard Turnbull, Harold Van Every, Larry Borsheim and Norman Grant. They all returned home and were successful in various aspects of business.

My wife, Lorraine, and I were granted three wonderful children, seven grandchildren and now two great-grandchildren. We are deeply grateful for the good that God has granted to us.

My original writing of this book began shortly after the war. For decades, it sat on a shelf waiting final editing. Now at long last, in my eighty-eighth year and at the insistence of my family and friends, I have retrieved both my records and my mother's to complete this book.

- Leonard J. Kovar

About the Author

Len Kovar was born and raised in Minneapolis, Minnesota. He graduated from Roosevelt High School and, at the outbreak of World War II, he enlisted in the Army Air Corp. After training, he became a Bombardier / Navigator and served in the 541st Bomb Group, the 727th Squadron of the 15th Air Force, which was located in Foggia, Italy. During his eleventh mission, Len Kovar was shot down, captured, and became a German Prisoner of War for nine months.

After the war, Len attended Macalester College in St. Paul, Minnesota and received a B.A. Degree in Economics. There he met his future wife, Lorraine Bakke. He worked for the University of Minnesota as a Concert and Lecture Advisor before going back to school. Three years later, he graduated from Andover Newton Theological School in Boston where he earned an S.T.M. degree. Over the years he served churches in Missouri, Montana, Massachusetts, New Zealand and California.

Throughout his life, Len completed further graduate work in the field of counseling utilizing the insights of Carl Jung and Roberto Assagioli. He is a licensed Marriage, Family and Child Counselor in the State of California.

He continues to be active in speaking engagement and study groups. He has earned the rank of Third Degree Black Belt in Kenpo Karate. As of the first publication of this book, he and his wife of sixty-two years are retired and live in Sacramento, California. They have three children, seven grandchildren, and two great-grandchildren.

About the Pictures

Many of the pictures included in this book were taken by a fellow kriege with a hidden camera. While in the POW camps, I signed up to receive copies of his pictures. True to his word, he sent these pictures shortly after the war ended. The photos capture scenes in which I was involved. All of these pictures are marked with a ' * '.

Also included in the book are pictures of my crew, especially Rick Turnbull and Mitch Cohen, friends and fellow crew-members. We were together in combat, and also on the ground as prisoners in a number of very dangerous situations. They survived the war and returned home. Both are gone now but I remember them as brave and loyal men.

One final note, I have included pictures of documents you may find interesting.

Kovar, Hobbs, and Beczkalo: February 5, 1944

Back Row: Goekmeyer, Roach, Boebeck, Cohen, Brittian, Lynch
Front Row: Kovar, Ash, Gould, Turnbull

The Men of Len's Crew

Rick Turnbull and Len Kovar

Mitch Cohen, Len Kovar, Rick Turnbull reunited decades later

This is a defensive flack field that got very close to our plane. We bounced around quite a bit and lost one engine over the target when we were hit by some of this flack. On this day, we limped home alone. This picture was taken on August 17th, 1944 (a previous mission) as we attacked the oil refineries at Ploesti.

I include this picture here to show what flying through flack looks like.

Len Kovar

R E S T R I C T E D

HEADQUARTERS
FIFTEENTH AIR FORCE
APO   520

C-UPD-bmr

GENERAL ORDERS)
NUMBER   3757)

2 October 1944

Citation of Unit I

SECTION I -- CITATION OF UNIT

    Under the provisions of Circular No. 333, War Department, 1943, and Circular No. 89, Headquarters NATOUSA, 10 July 1944, the following unit is cited for outstanding performance of duty in armed conflict with the enemy.

    451ST BOMBARDMENT GROUP. For outstanding performance of duty in armed conflict with the enemy. Notified to prepare their aircraft for a vital mission against the Markersdorf Airdrome, Vienna, Austria, in a counter air operation, the ground crews worked with enthusiasm to insure the mechanical perfection of their planes for the forthcoming mission. On 23 August 1944, twenty-four (24) B-24 type aircraft, heavily loaded with maximum tonnage, took off and set course for their destination. Enroute the formation was intercepted by numerous enemy fighters in a well coordinated attack, emerging from protective cloud covering six (6) to ten (10) abreast and employing twenty-millimeter cannon in their violent assaults. The highly aggressive enemy fighters made suicidal attempts against the bombers, in a desperate attempt to break up and destroy the formation, to prevent the successful completion of their vital mission. Displaying outstanding courage, professional skill and fortitude, the gallant crews battled their way through the overwhelming enemy opposition to the target, where, under continued heavy opposition, they completed a highly successful bombing run. Through their superior ability to maintain a tight protective formation and to direct heavy defensive fire against the fierce attacks of the enemy, the Group accounted for twenty-nine (29) enemy aircraft destroyed or damaged in the air. The excellent bombing pattern on the ground installations inflicted grave damage to important buildings and supplies, and twelve (12) enemy planes were destroyed on the ground. Throughout the aerial battle, the 451st Bombardment Group lost nine (9) heavy bombers, with others severely damaged by heavy enemy fire. Through this outstanding achievement a telling blow was struck at the fighter aircraft concentrations in the Vienna area, thus effectively and seriously crippling enemy operational efficiency at a time of great importance. By the outstanding courage, professional skill and unwavering determination of the combat crews, together with the superior technical skill and devotion to duty of the ground personnel, the 451st Bombardment Group has upheld the highest traditions of the Military Service, thereby reflecting great credit upon themselves and the Armed Forces of the United States of America.

By command of Major General TWINING:

R.K. Taylor,
Colonel, GSC
Chief of Staff.

OFFICIAL:

/s/ J. M. Ivins
J. M. IVINS,
Colonel, AGD,
Adjutant General.

A TRUE COPY:

Lynn J. Bartlett Jr.
LYNN J. BARTLETT, JR.,
Captain, Air Corps.

DISTRIBUTION: "D"

Letter of Citation for Len's Unit

WWII Prisoner of War - How I Survived

This is a previous mission on the way to a target.

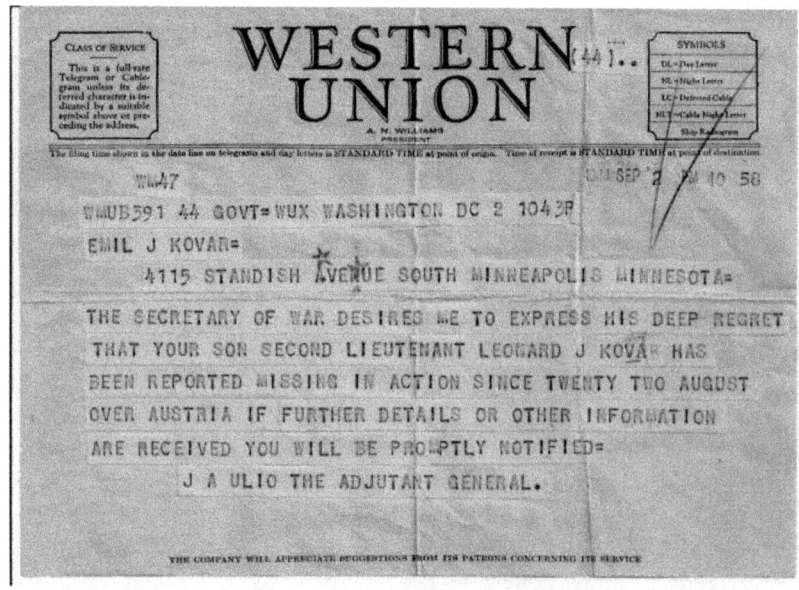

Telegram sent to my parents informing them I was MIA

Aug 31, 1944

Gosh! Dad, and Mom, and Sis, and Little Seno too! I heard the swellest news today, the very best that has ever come out of this war, makes me so happy I could cry. Now that I know there is hope, what more could I ask for, than to put that hope in "Him Above" and wait. What care I if the days go slowly, the weeks drag, and the months never end as long as that "hope" is fulfilled. Won't you all please join me in that waiting.

I'm sure the War Department has notified you that Len. went down. This afternoon I went up to his Group to talk to

This three-page letter was written to my parents by my friend, Bill Beczkalo, a fellow bombardier (in another plane). He flew in the same mission when I was shot down, but he made it back to base.

the fellows who were on that flight. It's a fact that "ten" beautiful white chutes came out of that ship. One of those was Ben. Even now he may be making his way back through enemy lines or else he might be a Prisoner of War. In either case it may be a month or two before any news gets out. The way the headlines look, it may be even sooner.

I know where and why their ship went down but it's better I should leave that out. Was talking to a fellow who has completed his missions and should be home in Mpls. in about three weeks. He

Page 2 of a 3-page letter to my parents. Bill also arranged to have my uniforms and personal items sent home.

III

has all the information and is going to drop in to see you all. I'm very sorry but I've forgotten his name but he will introduce himself – he is a grand guy.

Will do my utmost to get all the information possible and relay it to you as time goes by.

Best wishes to you all. I'm not much at this literary stuff but if you like I'll write often. Hoping to hear from you.

Your friend
Bill

Page 3 of a 3-page letter Bill wrote to my parents. His kindness meant the world to my parents in a time of great worry and stress.

WWII Prisoner of War - How I Survived

> **WAR DEPARTMENT**
> THE ADJUTANT GENERAL'S OFFICE
> WASHINGTON 25, D.C.
>
> IN REPLY REFER TO
> AG 201 Kovar, Leonard J.
> PC-N HAMMOND
>
> 5 September 1944
>
> Mr. Emil J. Kovar
> 4115 Standish Avenue, South
> Minneapolis, Minnesota
>
> Dear Mr. Kovar:
>
> This letter is to confirm my recent telegram in which you were regretfully informed that your son, Second Lieutenant Leonard J. Kovar, O-706317, Air Corps, has been reported missing in action over Austria since 22 August 1944.
>
> I know that added distress is caused by failure to receive more information or details. Therefore, I wish to assure you that at any time additional information is received it will be transmitted to you without delay, and, if in the meantime no additional information is received, I will again communicate with you at the expiration of three months. Also, it is the policy of the Commanding General of the Army Air Forces upon receipt of the "Missing Air Crew Report" to convey to you any details that might be contained in that report.
>
> The term "missing in action" is used only to indicate that the whereabouts or status of an individual is not immediately known. It is not intended to convey the impression that the case is closed. I wish to emphasize that every effort is exerted continuously to clear up the status of our personnel. Under war conditions this is a difficult task as you must readily realize. Experience has shown that many persons reported missing in action are subsequently reported as prisoners of war, but as this information is furnished by countries with which we are at war, the War Department is helpless to expedite such reports. However, in order to relieve financial worry, Congress has enacted legislation which continues in force the pay, allowances and allotments to dependents of personnel being carried in a missing status.
>
> Permit me to extend to you my heartfelt sympathy during this period of uncertainty.
>
> Sincerely yours,
>
> *J. A. ULIO*
> Major General,
> The Adjutant General.

A letter sent to my parents defining MIA and assuring them that if any new information is obtained, they will be informed

Len Kovar

WAR DEPARTMENT
THE ADJUTANT GENERAL'S OFFICE
WASHINGTON 25, D. C.

IN REPLY REFER TO
AGPC-G 201 Kovar, Leonard J.
NAT 214 O-706317

12 September 1944.

Mr. Emil J. Kovar,
 4115 Standish Avenue, South,
  Minneapolis, Minnesota.

Dear Mr. Kovar:

  Reference is made to my telegram and letter of 2 September and 5 September, respectively, in which you were informed that your son, Second Lieutenant Leonard J. Kovar, O-706317, has been missing in action since 22 August 1944.

  A report has now been received that he became missing in action over Hungary instead of over Austria.

  You may be assured that immediately upon receipt of any report concerning your son, it will be conveyed to you promptly.

      Sincerely yours,

      J. A. ULIO
      Major General,
      The Adjutant General.

A letter sent to my parents updating them on new information

WWII Prisoner of War - How I Survived

FIFTEENTH AIR FORCE
Office of the Commanding General
A P O 520

15 September 1944

Mr. Emil J. Kovar
4115 Standish Avenue
South Minneapolis, Minnesota

Dear Mr. Kovar:

The Liberator on which your son, Second Lieutenant Leonard J. Kovar, O-705317, was the bombardier failed to return from a combat mission to Vienna, Austria, on August 22, 1944. Since that date he and his crew are missing in action.

The facts available here are too indefinite as the basis for an opinion on your son's safety. Enemy fighters disabled Leonard's plane in the vicinity of Lebeny, Hungary. Several parachutes were seen leaving the damaged aircraft and it is believed that the entire crew bailed out. I wish it were possible for me to give you more definite details but I can not do so in the absence of supporting facts. You may be sure that you will be advised promptly if any additional information about your son is received.

Leonard's absence is keenly felt by all his fellow airmen. He was respected and admired by all who knew him for his courage and loyalty. I am proud to have had him in my command.

Very sincerely yours,

N. F. TWINING
Major General, USA
Commanding

True to their word, this letter was sent to my parents updating them on new information

HEADQUARTERS, ARMY AIR FORCES
WASHINGTON

AAF 201 - (8000) Kovar, Leonard J.
0706317

September 28, 1944.

Mr. Emil J. Kovar,
4115 Standish Avenue South,
Minneapolis, Minnesota.

Dear Mr. Kovar:

I am writing you with reference to your son, Second Lieutenant Leonard J. Kovar, who was reported by The Adjutant General as missing in action over Hungary since August 22nd.

Further information has been received which indicates that Lieutenant Kovar was a crew member of a B-24, (Liberator), bomber which departed from Italy on a combat mission to Vienna, Austria on August 22nd. Full details are not available, but the report indicates that during this mission while enroute to the target, at about 10:10 a.m., over Labany, Hungary, our planes encountered hostile aircraft and in the ensuing engagement your son's bomber sustained damage. Subsequently, the disabled craft dropped out of formation and all of the crew members were observed to parachute before it disappeared from sight. The report further indicates that the crew members of accompanying planes returning from the mission were unable to furnish any other details relative to Lieutenant Kovar's disappearance, therefore these facts constitute all the information presently obtainable.

Due to necessity for military security, it is regretted that the names of those who were in the plane and the names and addresses of their next of kin may not be furnished at the present time.

Please be assured that a continuing search by land, sea, and air is being made to discover the whereabouts of our missing personnel. As our armies advance over enemy occupied territory, special troops are assigned to this task, and all agencies of the government in every country are constantly sending in details which aid us in bringing additional information to you.

Very sincerely,

C. A. Oakley

C. A. OAKLEY,
First Lieutenant, A. C.,
Notification Branch,
Personal Affairs Division,
Assistant Chief of Air Staff, Personnel.

And information continued to be sent as it became available

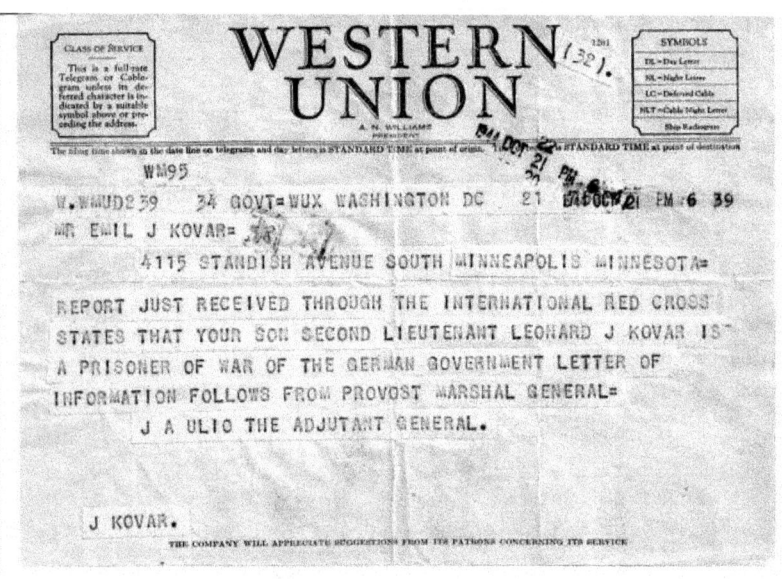

The telegram sent to my parents informing them that I was a POW - no longer missing in action

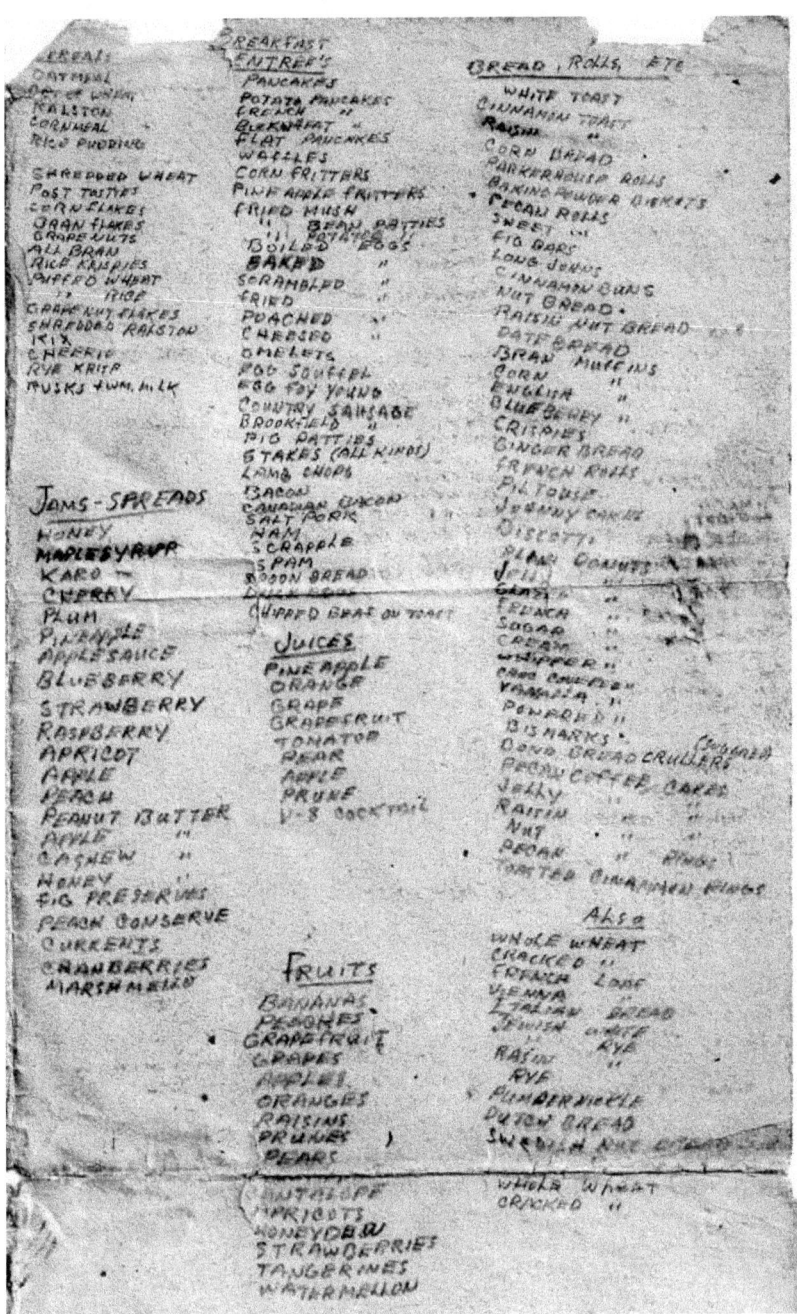

My precious list of "dream foods" (well-worn, often folded & kept close)

WWII Prisoner of War - How I Survived

```
MEATS
ROAST BEEF
  "   LAMB
  "   PORK
  "   CHICKEN
  "   TURKEY
  "   DUCK
  "   PHEASANT
  "   VENISON
LIVER - BOILED & FRIED
HAMBURGER
TENDERLOIN STEAK (M
T BONE       "
ROUND        "
CLUB         "
SIRLOIN      "
PORTERHOUSE  "
PORK CHOPS
VEAL   "
SPARE RIBS
HAM
SWEDISH SAUSAGE
GIBLETS
KIDNEYS
SHOULDER
LEG OF LAMB
HAM HOCKS
TONGUE
```

Close Up - Meats

```
SEA FOODS
OYSTERS - FRIED
   "     STEWED
   "     SOUP
   "     DRESSING
   "     RAW
SHRIMP   BOILED
   "     FRIED
   "     DRESSING
         COCKTAIL
LOBSTER
TUNA
TROUT - BROILED
   "    FRIED
   "    BAKED
        FRESH WATER
SHEEP HEAD
RED FISH
FLOUNDER
RED SNAPPER
MACKEREL
SALMON
CLAMS
HALIBUT
SWORDFISH
MACKEREL
HERRING
SMOKED CISCO
```

Close Up - Seafood

```
BUTTER BRICKEL
GRAHAM CRACKER
KREML
PORCUPINE PUDDING
DEVIL FOOD CAKE - PLAIN HERSHEY - COVER WITH CHOCOL
                                                  PU
FRIED BANANAS
  "   PINEAPPLE

BLACK BOTTOM PIE: - CHOCOLATE FILLING - COVER WITH
                   VANILLA WAFERS - CREAM FILLING
PORCUPINE PUDDING - SPONGE CAKE SATURATED WITH FR
JUICE (ORANGE) - STICK IN PECANS - TOP WITH
WHIPPED CREAM & CHERRIES
```

Close Up - Desserts

ADDRESS REPLY TO
COMMANDING GENERAL, ARMY AIR FORCES
WASHINGTON 25, D. C.

ATTENTION: AFPPA-8

HEADQUARTERS, ARMY AIR FORCES
WASHINGTON

(8000) Kovar, Leonard J.
O706317

December 9, 1944.

Mr. Emil J. Kovar,
4118 Standish Avenue South,
Minneapolis, Minnesota.

Dear Mr. Kovar:

    For reasons of military security it has been necessary to withhold the names of the air crew members who were serving with your son at the time he was reported missing.

    Since it is now permissible to release this information, we are inclosing a complete list of names of the crew members.

    The names and addresses of the next of kin of the men are also given in the belief that you may desire to correspond with them.

    Sincerely yours,

    Clyde V. Pinter

CLYDE V. PINTER
Colonel, Air Corps
Chief, Personal Affairs Division
Assistant Chief of Air Staff, Personnel

1 Incl
  List of crew members & names
  & addresses of next of kin

My parents received this letter along with a list of names and contact information for the other members of my crew. It means a lot to the families of missing military personnel to be able to talk to others in the same situation.

WWII Prisoner of War - How I Survived

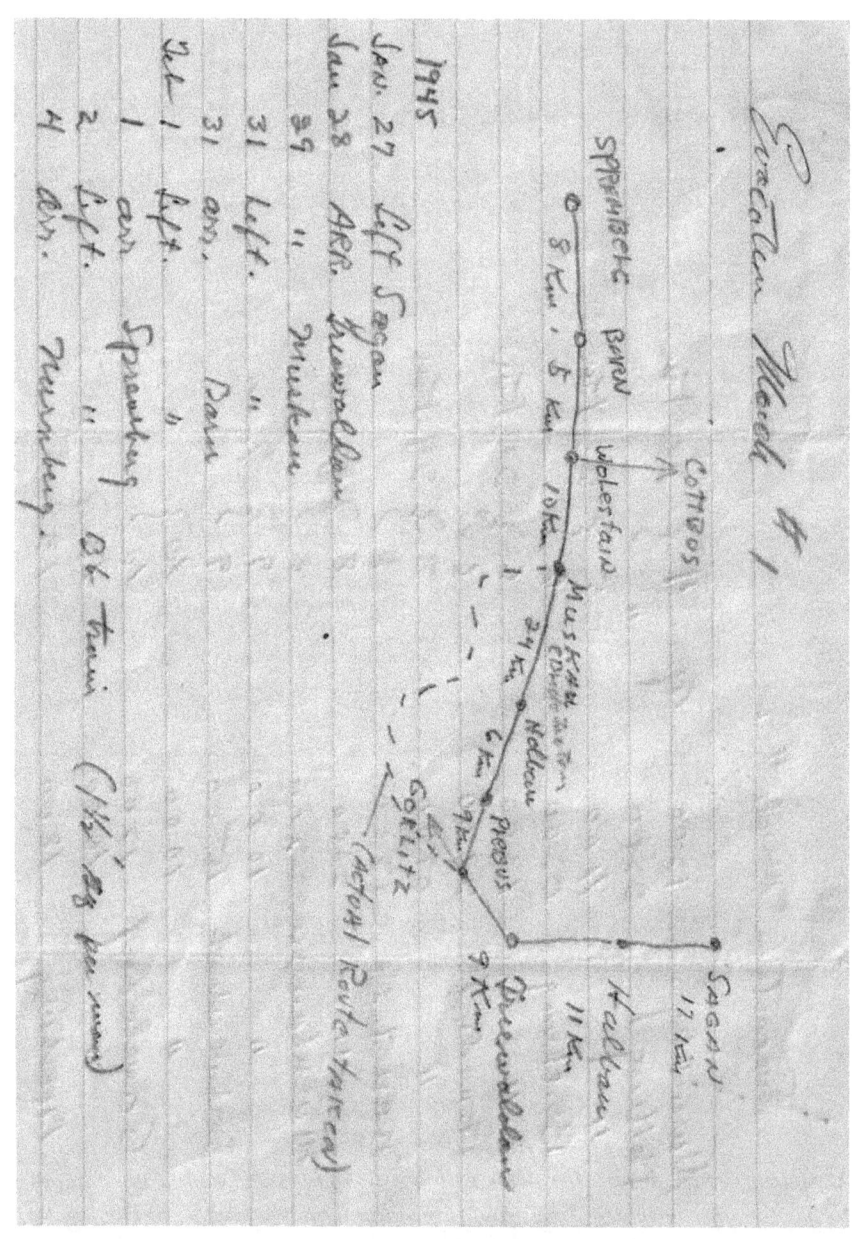

Notes On the First March

Notes On the Second March

www.ingramcontent.com/pod-product-compliance
Lightning Source LLC
Chambersburg PA
CBHW071709090426
42738CB00009B/1719